THE BILINGUAL SERIES OF
THE MOST IMPRESSIVE BEAUTY OF CHINA

饮食文化

CUISINE CULTURE

主　编◎青　闰
副主编◎秦　琴　王艳玲
参　编◎张连亮　张喜云　郑　盈

中国科学技术大学出版社

内容简介

"最美中国双语系列"是一套精品文化推广图书,包括《风景名胜》《民俗文化》《饮食文化》《杰出人物》《科技成就》《中国故事》六册,旨在传播中华优秀文化,传承中华民族宝贵的民族精神,展示奋进中的最美中国,可供广大中华文化爱好者、英语学习者及外国友人参考使用。

本书介绍了中国独具特色的饮食文化与部分代表性美食,展示了中国饮食文化的多姿多彩与深厚底蕴。

图书在版编目(CIP)数据

饮食文化:英汉对照/青闰主编.—合肥:中国科学技术大学出版社,2021.11
(最美中国双语系列)
ISBN 978-7-312-05208-8

Ⅰ.饮… Ⅱ.青… Ⅲ.饮食—文化—中国—英、汉 Ⅳ.TS971.2

中国版本图书馆CIP数据核字(2021)第121014号

饮食文化
YINSHI WENHUA

出版	中国科学技术大学出版社 安徽省合肥市金寨路96号,230026 http://press.ustc.edu.cn https://zgkxjsdxcbs.tmall.com
印刷	安徽国文彩印有限公司
发行	中国科学技术大学出版社
经销	全国新华书店
开本	880 mm×1230 mm 1/32
印张	7.75
字数	187千
版次	2021年11月第1版
印次	2021年11月第1次印刷
定价	35.00元

前 言 Preface

文化是一个国家与民族的灵魂。"最美中国双语系列"旨在弘扬和推广中华优秀文化,突出文化鲜活主题,彰显文化核心理念,挖掘文化内在元素,拓展文化宽广视野,为广大读者了解、体验和传播中华文化精髓提供全新的视角。本系列图书秉持全面、凝练、准确、实用、自然、流畅的撰写原则,全方位、多层面、多角度地展现中华文化的源远流长和博大精深,对于全民文化素质的提升具有独特的现实意义,同时也为世界文化的互联互通提供必要的借鉴和可靠的参考。

"最美中国双语系列"包括《风景名胜》《民俗文化》《饮食文化》《杰出人物》《科技成就》《中国故事》六册,每册中的各篇文章以文化剪影为主线,以佳句点睛、情景对话和生词注解为副线,别出心裁,精彩呈现中华文化的方方面面。

"最美中国双语系列"充分体现以读者为中心的编写理念,从文化剪影到生词注解,读者可由简及繁、由繁及精、由精及思地感知中华文化的独特魅力。书中的主线和副线是一体两面的有机结合,不可分割,如果说主线是灵魂,副线则是灵魂的眼睛。

"最美中国双语系列"的推出,是讲好中国故事、展现中国立场、传播中国文化的一道盛宴,读者可以从中感悟生活。

《饮食文化》包括饮食文化概述、八大菜系和特色美食三大部分,这里有底蕴深厚的酒文化、茶文化,有色香味俱全的川菜、鲁菜、粤菜、湘菜、苏菜、浙菜、闽菜、徽菜,还有特色鲜明的北京烤鸭、冰糖葫芦、耳朵眼炸糕、肉夹馍、羊肉泡馍、臊子面、胡辣汤、刀削面、太谷饼、兰州牛肉拉面,更有令人垂涎的热干面、南翔小笼包、蟹壳黄烧饼、糖油粑粑、桂林米粉、螺蛳粉、过桥米线、鲜花饼……

本书由河南理工大学秦琴撰写初稿,焦作大学张连亮与王艳玲撰写二稿,中南林业科技大学郑盈、焦作大学张喜云撰写三稿,焦作大学青闰负责全书统稿与定稿。

最后,在本书即将付梓之际,衷心感谢中国科学技术大学出版社的大力支持,感谢朋友们的一路陪伴,感谢家人们始终不渝的鼓励和支持。

青 闰

2021年3月6日

目 录 Contents

前言 Preface ……………………………………………………… i

第一部分　饮食文化概述
Part Ⅰ　Overview of Cuisine Culture

中国美食　Chinese Cuisine ………………………………… 003
饮食习惯　Dietary Habits …………………………………… 008
餐桌礼仪　Table Manners …………………………………… 013
酒文化　　Wine Culture ……………………………………… 018
茶文化　　Tea Culture ……………………………………… 023

第二部分　八大菜系
Part Ⅱ　Eight Cuisines

川菜　Sichuan Cuisine ……………………………………… 031
鲁菜　Shandong Cuisine …………………………………… 036
粤菜　Guangdong Cuisine ………………………………… 041
湘菜　Hunan Cuisine ……………………………………… 046
苏菜　Jiangsu Cuisine ……………………………………… 051

浙菜 Zhejiang Cuisine ……………………………………………056

闽菜 Fujian Cuisine ………………………………………………061

徽菜 Anhui Cuisine ………………………………………………066

第三部分　特色美食
Part Ⅲ　Special Cuisines

北京烤鸭　Beijing Roast Duck …………………………………073

老北京炸酱面　Traditional Beijing Noodles with Soybean Paste

………………………………………………………………………078

冰糖葫芦　Candied Haws on a Stick ……………………………083

驴打滚儿　Glutinous Rice Rolls Stuffed with Red Bean Paste

………………………………………………………………………088

狗不理包子　Go Believe …………………………………………093

耳朵眼炸糕　Earhole Fried Cake ………………………………097

桂发祥十八街麻花　Guifaxiang 18th Street Fried Dough Twists

………………………………………………………………………102

德州扒鸡　Dezhou Braised Chicken ……………………………107

肉夹馍　Marinated Meat in Baked Bun …………………………112

凉皮　Steamed Cold Rice Noodles ………………………………117

羊肉泡馍　Pita Bread Soaked in Lamb Soup …………………122

臊子面　Saozi Minced Noodles …………………………………127

胡辣汤　Hot Spicy Soup …………………………………………132

刀削面　Sliced Noodles …………………………………………137

目录

太原头脑　Soup with Eight Ingredients ……………………142

莜面栲栳栳　Hulls Oats Flour Kaolaolao Noodles …………147

太谷饼　Taigu Cake …………………………………………152

兰州牛肉拉面　Lanzhou Beef Ramen ……………………157

热干面　Hot-dry Noodles …………………………………162

精武鸭脖　Jingwu Duck-neck ………………………………167

南翔小笼包　Nanxiang Small Steamed Buns ……………172

生煎包　Pan-fried Baozi Stuffed with Pork ………………177

蟹壳黄烧饼　Xiekehuang Sesame Seed Cake ……………182

长沙臭豆腐　Changsha Preserved Smelly Tofu …………188

口味虾　Spicy Crawfish ……………………………………193

糖油粑粑　Sugar Oil Baba …………………………………199

桂林米粉　Guilin Rice Noodles ……………………………204

螺蛳粉　Liuzhou River Snails Rice Noodles ………………209

过桥米线　Crossing Bridge Rice Noodles …………………214

鲜花饼　Flower Cake ………………………………………219

粤式早茶　Guangdong Morning Tea ………………………224

煲仔饭　Steamed Rice in Clay Pot …………………………229

云吞面　Wonton Noodles …………………………………234

第一部分 饮食文化概述

Part I Overview of Cuisine Culture

饮食文化概述 第一部分

中国美食

Chinese Cuisine

 导入语　Lead-in

中国饮食讲究色香味形俱全,民间俗话说:"民以食为天,食以味为先。"从南到北,从东到西,生活在不同地区的中国人享受着丰富多样的食物。由于各地食材不尽相同,因此中国美食凭借所用的不同食材形成了各具特色的美食。不同的地域食材又演变出不同的烹饪方法,中国烹饪风格以川菜、鲁菜、粤菜、湘菜、苏菜、浙菜、闽菜、徽菜八大菜系为代表。中国饮食文化绵延三千一百余年,历经生食、熟食、药膳养生、自然烹饪和科学烹饪五个阶段,推出六万多种传统菜点,享有"美食王国"的美誉。中国饮食文化既影响了朝鲜、韩国、泰国、新加坡、日本、蒙古等亚洲国家,也辐射到了欧洲、美洲、非洲和大洋洲。

 文化剪影　**Cultural Outline**

Chinese cuisine is the general term of the dishes of various regions in China, which has rich **connotations**① including food resources, cutting and cooking skills. Therefore, it is the **crystallization**② of thousands of years' history of Chinese cooking and the wisdom of all Chinese nationalities.

中国美食是中国各地菜系的总称,具有丰富的内涵,包括食物、刀工和烹饪技巧。因此,它是中国几千年烹饪历史和各民族智慧的结晶。

Chinese cuisine can be cooked in different techniques. There're over thirty cooking techniques in Chinese cuisine, such as frying, deepfrying, smoking, roasting, mixing, boiling and steaming, and so on.

中国美食可以用不同的烹饪方法,现有煎、炸、熏、烤、拌、水煮和清蒸等三十多种烹饪方法。

Seasoning③ plays a vital role in Chinese cooking. There're over one hundred kinds of **spices**④ that are now used as seasoning in Chinese food. People add seasoning to food at the right time with the right mount, resulting in the desired flavor.

调料在中国烹饪中也扮演着至关重要的角色。目前,中餐中用于调味的香料有一百多种。在适当的时间加上适量的调料,人们就能得到想要的味道。

佳句点睛 Punchlines

1. For Chinese, taste is a matter of not only tongue and nose, but also the heart.

对中国人来说,味道不仅关乎舌鼻,而且关乎心灵。

2. Tastes of Chinese food are traditionally **categorized**⑤ into five tastes: salty, **spicy**⑥, vinegary, sweet and bitter.

中国食物的口味传统上分为咸、辣、酸、甜和苦五种。

3. Chinese cuisine is not only about our life, but also closely linked with the Chinese culture.

中国饮食不仅关乎我们的生活,而且与中华文化息息相关。

情景对话 Situational Dialogue

A: Hi, where are you going?

B: I'm going to KFC. Do you need to bring you a hamburger?

A: No, thank you. Obviously, the current traditional Chinese food is **overshadowed**⑦ by fast food nowadays.

B: In such a fast-paced life, eating a delicious meal is a very **luxurious**⑧ enjoyment. Therefore, many people have chosen fast food as a convenient and fast way to eat. No doubt it meets the exact needs of modern people.

A: But fast food is high calorie, which is bad for our health.

B: Not all, many fast foods have made great changes, they invented some dishes according to Chinese people's taste and flavor. We can enjoy a better **nutritional**⑨ structure and choose our favorite food.

A: I don't think so. I think our Chinese food is so varied and beautiful that it is worth choosing a lot.

B: Really?

A: If you don't believe me, I'll take you to sample some of our **authentic**⑩ Chinese cuisine. You'll be sure to have a good time and see what the real "Kingdom of Cuisine" is.

B: Ok, let's go then.

A: 嗨,你要去哪里?

B: 我要去肯德基,需要给你带个汉堡包吗?

A: 不用了,谢谢。现在,中国的传统食物显然受到了快餐的极大冲击。

B: 在如此快节奏的生活中,坐下来吃一顿美味的饭菜是一种非常奢侈的享受。许多人选择快餐是因为它方便快捷。毫无疑问,它满足了现代人们的需求。

A: 但是,快餐热量很高,对我们的健康没有什么好处。

B: 不都是这样,很多快餐做出了极大的改变,他们根据中国人的口味创新出一些菜肴。我们可以根据自己喜欢的口味选择营养丰富的食物。

A: 我倒不这么认为。我觉得咱们中国的美食各式各样,美不胜收,值得一试。

B: 是吗?

A: 不信我带你去尝尝,到时候你一定会觉得不虚此行,明白什么是真正的"美食王国"。

B: 好啊,那咱们走吧。

生词注解 Notes

① connotation /ˌkɒnəˈteɪʃn/ *n.* 内涵;含蓄

② crystallization /ˌkrɪstəlaɪˈzeɪʃən/ *n.* 结晶化;具体化

③ seasoning /ˈsiːzənɪŋ/ *n.* 调味品;佐料

④ spice /spaɪs/ *n.* 香料;调味品

⑤ categorize /ˈkætəɡəraɪz/ *vt.* 将……分类;把……归类

⑥ spicy /ˈspaɪsɪ/ *adj.* 辛辣的;多香料的

⑦ overshadow /ˌəʊvəˈʃædəʊ/ *vt.* 使……黯然失色;使……相形见绌

⑧ luxurious /lʌɡˈʒʊərɪəs/ *adj.* 奢侈的;豪华的

⑨ nutritional /njuˈtrɪʃənl/ *adj.* 营养的;滋养的

⑩ authentic /ɔːˈθentɪk/ *adj.* 真正的;正宗的

饮食习惯

Dietary Habits

 导入语　Lead-in

随着社会的发展，人类的饮食结构也得到了改善。从采摘野果、捕猎到用火烤食，五谷杂粮逐渐成为人类的主食，鱼、肉、蛋、奶、蔬菜和水果在日常饮食中占据的比例逐渐增加。中国人饮食习俗的一大特点是以熟食为主，人们通过煎、炒、炸、煮、蒸、炖、烧、烤等方式，使食物呈现不同的形式。另一个显著特点是中国人保存食物的方式和数量比其他国家多，如种类繁多的腌制、熏制食品等。通过烟熏、盐腌、加糖、浸泡在各种调料中使食物得以保存，所选的食材包括谷物、肉类、鸡蛋、蔬菜等。

文化剪影　Cultural Outline

The Chinese diet values the choice of spices, the preparation of soup bases, and the careful combination of ingredients of different flavors and textures. These are the perfect **embodiment**① of the Chinese philosophy of balanced diet.

中国饮食重视香料选择、汤底准备以及不同口味、质地食材的精心组合。这些是中国饮食均衡哲学的完美体现。

China is a big agricultural country with a long history, mainly growing wheat in the north and rice in the south. Due to the different natural and social conditions and lifestyles, the **dietary**② habits are not the same.

中国是一个具有悠久历史的农业大国。中国北方主要种植小麦,南方主要种植水稻。由于自然条件、社会条件和生活方式的不同,各地的饮食习惯也不尽相同。

The Chinese dietary habits are **reflected**③ in their ideas and beliefs about food. Chinese people believe that food is also medicine, and they make various delicacies with the medicinal value of food materials.

中国人的饮食习惯体现在对食物的观念和信仰上。中国人认为食物也是药物,人们利用食物原料的药用价值,做出各种美味佳肴。

饮食文化

佳句点睛 Punchlines

1. Chinese way of eating is characterized by a notable **flexibility**④ and **adaptability**⑤.

中国人的饮食方式具有显著的灵活性和适应性。

2. Chinese cuisine forms the profound Chinese food culture, which is **inseparable**⑥ from Chinese traditional festivals.

中国美食形成了深厚的中国饮食文化,这与中国的传统节日密不可分。

3. With the progress of science and technology, people pay more attention to the **rationality**⑦ of diet while meeting their physiological needs the pursuit of health has increased day by day.

随着科技的进步,人们在满足生理需求的同时更加注重饮食的合理性,对健康的追求也与日俱增。

情景对话 Situational Dialogue

A: I'm lucky to see you at our school's **canteen**⑧.

B: I'm also glad to see you.

A: Can we have lunch together?

B: Sure. Let's have a seat.

A: What do you like to have every day?

B: In the morning, I'd like to eat porridge and fried dough sticks. At lunch, a bowl of noodles is my favorite. You know, we have different kinds of noodles in China.

A: Oh, the Chinese food is totally different from ours. I would choose milk in the morning and some meat at lunch. From the perspective of different dietary habits, the East is given priority to **vegetarian**[9] food while the West is given priority to meat and milk.

B: I can't agree with you anymore.

A: It's time to attend the class. I'm so pleased to talk with you. Have a nice day!

B: Have a nice day!

A: 很幸运在我们学校的食堂见到你。

B: 我也很高兴见到你。

A: 我们能一起吃午饭吗?

B: 当然,我们找个座位坐下吧。

A: 你每天都喜欢吃什么?

B: 早上我喜欢喝粥、吃油条。午餐最喜欢吃面条。你知道,在中国我们有各种各样的面条。

A: 噢,中国菜和我们国家的完全不一样。我们早上喝牛奶,中午会吃些肉。从不同的饮食习惯来看,东方以素食为主,西方以肉类和牛奶为主。

B: 我完全同意你的观点。

A: 该上课了。很高兴和你聊天,祝你今天过得愉快!

B: 也祝你愉快！

 生词注解 Notes

① embodiment /ɪmˈbɒdɪmənt/ n. 体现；化身

② dietary /ˈdaɪətərɪ/ adj. 饮食的；规定食物的

③ reflect /rɪˈflekt/ vt. 反映；反射

④ flexibility /ˌfleksəˈbɪlətɪ/ n. 灵活性；弹性

⑤ adaptability /əˌdæptəˈbɪlətɪ/ n. 适应性；可变性

⑥ inseparable /ɪnˈseprəbl/ adj. 不可分的；不愿分开的

⑦ rationality /ˌræʃəˈnælətɪ/ n. 合理性；合理的行动

⑧ canteen /kænˈtiːn/ n. (工厂、商店、高校的)食堂

⑨ vegetarian /ˌvedʒəˈteərɪən/ adj. 素食的；素的

饮食文化概述 第一部分

餐桌礼仪

Table Manners

导入语 Lead-in

饮食文化是中华民族独具特色的文化之一,也是研究中国文化的一把钥匙。中国享有"礼仪之邦"的美誉,饮食礼仪是展现社会文明的窗口,个人修养和礼仪礼节在餐桌上可以得到全面展现。餐桌礼仪在中国人的生活秩序中占有非常重要的地位。每个民族在长期的社会实践中都有自己的一套规范化的饮食礼仪,每个社会成员都应该自觉遵守行为准则。如果能在宴饮时保持良好的进餐形象和礼仪修养,不仅可以得到他人尊重,而且可以促进自身社交和事业发展。但是,必须注意宗教饮食禁忌,尤其是主宾的饮食禁忌。

饮食文化

文化剪影 Cultural Outline

Table manners, including the choice of the place of the **banquet**①, the dining environment, the preparation of **tableware**②, the way of treating guests and the banquet row seats, are reflected a respect to the guests.

餐桌礼仪包括宴请场所的选择、环境的营造、餐具的准备、待客的方式、宴席的排座等,都体现了对客人的一种尊重。

Chinese people show great **hospitality**③ at the dinner table. For example, Chinese always like sitting around the table to eat and drink together, speak and laugh loudly to create warm table atmosphere.

中国人在餐桌上都非常殷勤好客。比如,中国人总是喜欢围坐在餐桌旁一起吃喝、大声说笑,以营造温暖的餐桌气氛。

Chopsticks④ are the most important tableware for Chinese. Before dinner, chopsticks should be put on the right of the bowl tidily. After meal, it should be placed in the middle of the bowl. When talking with other people, we'd better put down the chopsticks.

筷子是中国人最重要的餐具。饭前,筷子应该整齐地放在饭碗右侧。饭后,则摆在饭碗中间。和别人交谈时,最好放下筷子。

佳句点睛 Punchlines

1. Your cup should be lowered than others' when you **propose**⑤ a toast.

敬人酒水时杯口应低于对方杯口。

2. We'd better use public chopsticks when taking food for others.

为他人夹菜时最好使用公筷。

3. Answering the phone during the meal, we should go outside.

就餐时接听电话应到室外。

情景对话 Situational Dialogue

A: Hi, I'm gonna study abroad next month. Would you like to tell me something about the table manners in China?

B: Of course. The Westerners use plates as **containers**⑥ of food, but Chinese use and bowls as those of rice. Chopsticks and knives & forks are the most basic differences between Chinese and Western table manners.

A: Oh, it's a little **complicated**⑦ because I have never used chopsticks before.

B: It doesn't matter. I can guide you to use them.

A: You're so kind.

B: I'm glad to share with you the traditional Chinese culture. Look at me. Try to use them to pick up the peanuts.

A: Let me have a try. It's too hard for me to pick up the peanuts with chopsticks.

B: You need to practice. By the way, please remember some dining **taboos**⑧ when using chopsticks.

A: Taboos?

B: Yes, Chinese people usually don't beat their bowls while eating, since the behavior used to be practiced by beggars. Besides, don't point at people with chopsticks because it is impolite.

A: OK, I'll try my best to remember all the taboos.

B: It will be good for you. Good luck!

A: 嗨,下个月我要出国留学。你能给我讲讲中国的餐桌礼仪吗?

B: 当然可以。西方人用盘子盛食物,但中国人用碗盛饭。筷子和刀叉是中西餐桌礼仪最基本的区别。

A: 噢,有点儿复杂,因为我以前从来没有用过筷子。

B: 没关系,我可以教你。

A: 你真好。

B: 很高兴和你分享中国传统文化。看我,试着用筷子夹起花生。

A: 让我试试,这对我来说太难了。

B: 你需要练习。对了,使用筷子时还请记住一些用餐禁忌。

A: 禁忌?

B: 是的,使用筷子时有一些忌讳。中国人吃饭时,忌讳用筷子敲打饭碗,因为这像是一个乞丐乞讨时的行为。另外,不要用筷子指人,这样不礼貌。

A: 好的,我会努力记住的。

B: 这对你会有好处。祝你好运!

生词注解 Notes

① banquet /ˈbæŋkwɪt/ n. 宴会;盛宴

② tableware /ˈteɪblweə(r)/ n. 餐具;碗筷

③ hospitality /ˌhɒspɪˈtæləti/ n. 好客;殷勤

④ chopstick /ˈtʃɒpstɪk/ n. 单根筷子

⑤ propose /prəˈpəʊz/ vt. 建议;提议

⑥ container /kənˈteɪnə/ n. 容器;集装箱

⑦ complicated /ˈkɒmplɪkeɪtɪd/ adj. 难懂的;复杂的

⑧ taboo /təˈbuː/ n. 禁忌;禁忌语

酒文化

Wine Culture

  导入语　Lead-in

中国是世界上酒类品种最全、酿造历史最长、产业规模最大的国家。由于中国地域广阔，因此酒在中国也呈现多种多样的特点，其中以白酒最具代表性。中国白酒的蒸馏技术和酿造方法是中华民族的伟大发明，根据香型可分为酱香、清香、浓香、米香、凤香等。随着历史的发展，酒文化也逐渐演变升华，既包含酒本身的独特风格和意境，也包含人们饮酒时所形成的精神内涵和美妙感受。不少文人墨客在宴会、饯行时留下了许多经典的品酒、鉴酒佳话。

文化剪影 Cultural Outline

From literary creation, cultural entertainment, dietary cooking to healthcare and so on, wine plays an important role in Chinese people's life. Chinese **ancestors**① used wine to enjoy themselves while writing poetry, or make a **toast**② to their relatives and friends during a feast.

从文学创作、文化娱乐到饮食烹饪、养生保健等各方面,酒在中国人的生活中都占有重要位置。古代中国人在写诗时以酒助兴,或在宴会上向亲朋好友敬酒。

In the long history of human culture, wine is not only an **objective**③ material existence, but a cultural symbol. In the time-honored wine making and drinking, China have formed a unique wine culture.

在人类文化的历史长河中,酒已经不仅仅是一种客观的物质存在,而且是一种文化象征。在源远流长的酿酒和饮酒历史中,中国已经形成了独具特色的酒文化。

Wine can free people from fame and fortune, life pressure and social laws, **releasing**④ the natural of human nature and showing the heroic spirit of freedom. Many Chinese people like to express their feelings and talk about things warmed with wine.

酒可以让人从名利、生活压力和社会法则中解脱出来,尽情释放自然率真的人性本色,展现自由不羁的豪迈气概。很多中国人喜欢

在酒酣耳热后攀感情、谈事情。

 佳句点睛　Punchlines

1. Liquor is commonly used in society, and a good gift in communication.

白酒常用于交际，也是交往时的馈赠佳品。

2. The relationship between wine and **literature**⑤ has a long history in China.

在中国，酒和文学的关系源远流长。

3. Wine not only reflects the unique national customs, but also adds joy to people's life.

酒不仅能够体现独特的民族风俗，而且能给人们的生活增添欢乐。

 情景对话　Situational Dialogue

A: Do you know about wine culture?

B: I know something about it. Wine is a kind of special cultural **carrier**⑥, which plays a unique role in social **intercourse**⑦.

A: Can you tell me something about Chinese wine culture?

B: It's a long story. China is the hometown of wine, so wine culture not only has a long history in China, but also is an important part of Chinese national cuisine culture.

A: I'm a stranger here and want to listen more carefully.

B: Wine is not only a kind of drink, but also has spiritual and cultural values. As a spiritual culture, it is reflected in social and political life, literature and art and even **aesthetic**① taste. In this sense, drinking wine is also drinking culture.

A: What do you mean?

B: You know that from time immemorial, countless writers have left a lot of well-known poetic masterpieces through the ages related to wine, such as "From cups of jade that glow with wine of grapes at night, drinking to pipa songs, we are summoned to fight" "When will we have the bright moon? Holding up a wine cup, I ask the blue sky"...

A: It seems that the wine culture is really **extensive**② and profound, which makes people have endless aftertaste.

B: That's right. People not only taste the wine, but also enjoy the culture.

A：你了解酒文化吗？

B：我了解一些。酒是一种特殊的文化载体，在社交中占有独特的地位。

A：那你能给我讲讲中国的酒文化吗？

B：这个说来话长。中国是酒的故乡，酒文化不但在中国源远流长,而且是中华民族饮食文化的一个重要组成部分。

A:我初来乍到,想细听端详。

B:酒不仅是一种饮品,还具有精神文化价值。作为一种精神文化,它体现在社会政治生活、文学艺术乃至审美情趣等方面。从这个意义上说,饮酒也是一种文化。

A:此话何意?

B:你知道,古今无数文人墨客留下了许多与酒相关的千古绝唱,比如"葡萄美酒夜光杯,欲饮琵琶马上催""明月几时有,把酒问青天"……

A:看来酒文化真是博大精深,令人回味无穷啊。

B:说得没错。人们在觥筹交错中不仅品尝了美酒,也接受了文化洗礼。

生词注解 Notes

① ancestor /ˈænsestə(r)/ *n.* 祖先;始祖

② toast /təʊst/ *n.* 敬酒;祝酒

③ objective /əbˈdʒektɪv/ *adj.* 物体的;客观的

④ release /rɪˈliːs/ *vt.* 释放;松开

⑤ literature /ˈlɪtrətʃə(r)/ *n.* 文学;文献

⑥ carrier /ˈkærɪə(r)/ *n.* 载体;载波

⑦ intercourse /ˈɪntəkɔːs/ *n.* 交往;交流

⑧ aesthetic /iːsˈθetɪk/ *adj.* 审美的;具有审美趣味的

⑨ extensive /ɪkˈstensɪv/ *adj.* 广博的;广阔的

第一部分 饮食文化概述

茶文化

Tea Culture

导入语 Lead-in

中国是茶叶的原产国,以茶叶种植和制茶技艺闻名。《神农本草经》记载:"神农尝百草,日遇七十二毒,得茶而解之。"人们渐渐发现,喝茶不仅能治病,还能清热解渴,于是茶就成为了中国人生活中必不可少的一种饮品。中国茶种类繁多,主要有绿茶、黑茶、黄茶、青茶、红茶、白茶,以及经过熏制的花茶等。中国的饮茶习俗通过古老的"丝绸之路"和其他贸易渠道传播到了欧洲和其他许多地区。中华民族在世界文明史上书写了灿烂的茶文化篇章,而茶叶的开发和推广也成为中国对世界的重要贡献之一。

 文化剪影 Cultural Outline

The widespread **customs**① of growing, making and drinking tea in the world today are all spread out from China. Tea can not only satisfy people's physiological and material need, but the art of drinking tea can also make people get spiritual enjoyment and achieve a wonderful state.

当今世界广泛流传的种茶、制茶和饮茶习俗都是由中国向外传播出去的。茶既能满足人们的生理和物质的需求,茶艺又能使人们得到精神享受,达到一种美妙的境界。

The Classic of Tea provides a complete scientific basis for tea production, which introduces the history and development of tea in details. It is the earliest, most complete and comprehensive **monograph**② on tea in China and even the world, playing an active role in promoting the development of tea culture.

《茶经》为茶叶生产提供了完整的科学依据,该书详细介绍了茶叶的历史和发展,是中国乃至世界现存最早、最完整、最全面介绍茶的专著,对茶文化的发展起到了积极的推动作用。

Tea is closely connected with the spiritual world of the Chinese people to form the material basis of tea culture. Tea tasting has become a kind of **leisurely**③ lifestyle of the Chinese people, among which the function of self-cultivation of tea is more valued by people.

茶与中国人的精神世界息息相关，构成了茶文化的物质基础。品茶已经成为中国人的一种休闲生活方式，茶的修身养性的功能更为人们所重视。

佳句点睛 Punchlines

1. Tea once became the main exported commodity of China.
茶叶曾经是中国主要的出口商品。

2. The exploring, discovering, using and feeling tea by human beings is the history of tea culture.
人类探索、发现、使用、感受茶叶的历史，就是茶文化的历史。

3. The Chinese Tea Ceremony contains the wisdom of oriental **philosophy**④.
中国茶道蕴含着东方哲学的智慧。

情景对话 Situational Dialogue

A: Can I help you? What kind of tea would you like to buy?

B: I'm not familiar with tea. Would you like to introduce something for me?

A: OK, let's have a seat to enjoy a cup of tea.

B: Thank you.

A: Let's start from the original process. In the past, people either boiled the tea leaves straight from a tree, or sun-dried tea leaves for drinking. Later the processing methods have changed over the centuries, people only **oxidized**[5] it to **retain**[6] more of the original taste of the leaf very slightly.

B: Oh, but there're many kinds of tea, so I don't know how to choose.

A: The different kinds of tea can satisfy the different needs. Just choose it according to your flavor.

B: OK, I'd like to try different kinds of tea.

A: Wait for a moment. I'd like to prepare for the Chinese tea ceremony, so you can have a better understanding of Chinese tea.

B: That would be amazing. I heard Chinese tea ceremony reflect the core idea of **Taoism**[7], **Confucian**[8] and **Buddhism**[9], which is a combination of philosophy and lifestyle.

A: That's right. Here's some tea for you. Have a try, please.

B: Thanks a lot.

A：我能帮你吗？你想买什么茶呢？

B：我对茶不太了解。你能给我介绍一下吗？

A：好的，我们坐下来喝杯茶吧。

B：谢谢你。

A：让我们从头开始吧。过去，人们要么直接从树上采下茶叶煮着喝，要么把茶叶晒干后再泡水喝。茶叶的加工方法在后来的几个世纪中发生了变化,人们只是将茶叶稍加氧化,以保留更多原始的味道。

B:噢,但是茶叶的种类太多,我不知道怎么选。

A:不同的茶可以满足不同的需求,根据你的口味来选择。

B:好的,我想试试不同种类的茶。

A:稍等,我想展示一下中国的茶道,这样你就能更好地了解中国茶了。

B:太好了。我听说中国的茶道体现了儒、道、佛的核心思想,是哲学和生活方式的结合。

A:说得对。这是为你准备的茶,请品尝。

B:多谢。

生词注解 Notes

① customs /ˈkʌstəmz/　n. 风俗;海关

② monograph /ˈmɒnəɡrɑːf/　n. 专题著作;专题论文

③ leisurely /ˈleʒəlɪ/　adj. 空闲的;悠闲的

④ philosophy /fəˈlɒsəfɪ/　n. 哲学;哲理

⑤ oxidized /ˈɒksɪdaɪzd/　adj. 被氧化的;生锈的

⑥ retain /rɪˈteɪn/　vt. 保持;持有

⑦ Taoism /ˈdaʊɪzəm/　n. 道教;道家学说

⑧ Confucian /kənˈfjuːʃən/　adj. 儒家的;儒学的

⑨ Buddhism /ˈbʊdɪzəm/　n. 佛教;佛门

第二部分　八大菜系

Part II　Eight Cuisines

川菜

Sichuan Cuisine

导入语 Lead-in

四川菜简称"川菜",是中国传统八大菜系之一。川菜的历史悠久,最早起源于古代的巴国和蜀国,当地人喜欢大量使用花椒、姜等辛辣调料,这便是蜀人"尚滋味、好辛香"的起源。

川菜讲究色、香、味、形,在"味"字上格外下工夫,以味的多、广、厚著称。经过不断的融合发展,川菜的味道可进一步分为"麻、辣、咸、酸、甜、苦、香"七种味道和"干烧、酸辣、鱼香、干煸、怪味、椒麻、红油"八种滋味。"一菜一格,百菜百味"是人们对川菜的赞誉。川菜的招牌菜有麻婆豆腐、宫保鸡丁、回锅肉、夫妻肺片、毛血旺、东坡肘子、鱼香肉丝、干烧鳜鱼等。

饮食文化

文化剪影　Cultural Outline

Chilli was introduced into Sichuan for **cultivation**① and widely used in Sichuan cuisine, which was regarded as the beginning of the early **appearance**② of modern Sichuan cuisine.

辣椒被引进四川进行种植并广泛运用于川菜烹调中,被视为近代川菜初现雏形的开始。

Sichuan enjoys a reputation as "A Land of **Abundance**③," which **yields**④ rich cooking ingredients. The common cooking techniques of Sichuan cuisine are sautéing, frying, stir-frying without **stewing**⑤, dry-frying, homestyle frying and so on.

四川素有"天府之国"之称,物产富庶,烹饪用料广泛。川菜常用的烹调技法有小煎、小炒、干煸、干烧和家常烧等。

The introduction and wide **application**⑥ of **chilli**⑦ promote Sichuan cuisine to gradually form a system with extremely rich local flavor. Nowadays, Sichuan cuisine not only has a wide influence in China, but also has spread all over the world.

川菜逐步成为一个地方风味极其浓郁的体系,离不开辣椒的引进和广泛应用。如今,川菜不但在中国各地具有广泛的影响,而且已经遍及世界各地。

佳句点睛 **Punchlines**

1. Sichuan is rich in products, which provide rich raw materials for Sichuan cuisine.

四川物产丰饶,这些为川菜提供了丰富的原料。

2. Sichuan cuisine pays special attention to seasoning and is quite rich in flavor.

川菜特别注重调味,味型也相当丰富。

3. Sichuan cuisine can be carefully mixed into sour, sweet, bitter, spicy, tongue-numbing, **aromatic**® and salty flavors.

川菜的味道可进一步细分为酸、甜、苦、辣、麻、香、咸七种滋味。

 情景对话 **Situational Dialogue**

A: I've heard your early dream was to eat all over China. How about your plan?

B: There're eight cuisines in China, distributing throughout the different provinces. I have just tasted half of them.

A: Now what's your favorite dish?

B: I have to say that Sichuan cuisine is the best for people like me.

A: Is Sichuan food simply spicy?

饮食文化

B: In addition to the well-known spicy food, Sichuan cuisine is also characterized by its wide range of ingredients, various cooking methods and various tastes. Whoever you are, you can find the food that suits your taste.

A: Can you recommend some Sichuan specialties?

B: Would you like to have chicken or fish? There're spicy chicken and fish fillets in hot.

A: I'd like some vegetarian diet.

B: I guess Mapo Tofu is the best choice for you.

A: 听说你早年的梦想是吃遍中国。你计划得怎么样了？

B: 中国菜有八大菜系，分布于各省各地。我才品尝了一半。

A: 那最让你流连的是什么菜？

B: 不得不说我最爱川菜了。

A: 川菜只是单纯的辣吗？

B: 除了世人皆知的麻辣，川菜还具有取材广、做法多、口味多等特点，无论是谁，一定能在川菜里找到适合自己的口味。

A: 你能推荐几道特色川菜吗？

B: 你是想吃鸡还是鱼呢？鸡类有麻辣鸡，鱼类有水煮鱼等。

A: 我想吃点儿素的。

B: 那就来份麻婆豆腐吧。

生词注解 Notes

① cultivation /ˌkʌltɪˈveɪʃn/ n. 培养;耕作

② appearance /əˈpɪərəns/ n. 外貌;外观

③ abundance /əˈbʌndəns/ n. 充裕;丰富

④ yield /jiːld/ vt. 出产(产品或作物);产出

⑤ stewing /stjuːɪŋ/ n. 炖;蒸煮

⑥ application /ˌæplɪˈkeɪʃn/ n. 应用;申请

⑦ chilli /ˈtʃɪli/ n. 辣椒(经常在干后制成辣椒粉)

⑧ aromatic /ˌærəˈmætɪk/ adj. 芳香的;芬芳的

饮食文化

鲁菜

Shandong Cuisine

导入语 Lead-in

鲁菜又称为"山东菜",是中国八大菜系之一,主要由济南菜、胶东菜和孔府菜组成。鲁菜历史悠久,始于春秋战国时期的山东。明、清时期,鲁菜已是宫廷御膳的主体,随后又发展为喜庆寿诞的高档宴席和家常菜。山东地理位置优越、气候温和、河湖交错、物产富饶,为鲁菜提供了食材保障。人们习惯使用葱、姜、蒜提鲜,尽量保持食物本味。用"食不厌细,脍不厌精"来形容鲁菜最合适不过。鲁菜有葱烧海参、三丝鱼翅、白扒四宝和糖醋黄河鲤鱼等招牌菜。

文化剪影 Cultural Outline

After many years' sustainable development and evolution, Shandong cuisine has been continuously carried forward on the basis of **maintaining**① the traditional flavor together with the trend of the times.

鲁菜历经多年的可持续发展和演变,在保持传统风味的基础上,结合时代发展趋势,不断发扬光大。

Shandong cuisine is mainly characterized by salty and fresh flavor, proficient in cooking soup and **seafood**②. It is characterized by the cooking techniques include quick stir-frying, frying, stewing, boiling and braising.

鲁菜的主要特色以咸鲜为主,精于制汤,善烹海味。鲁菜以爆、炒、烧、塌、扒等烹调技巧最具特色。

Shandong is an important vegetable producing area in China because of its wide variety of high-quality vegetable. Moreover, the people of Shandong Province are simple and **good-hearted**③, and pay special attention to hospitality.

山东是中国重要的蔬菜产地,其种植的蔬菜种类繁多、品质优良。此外,山东民风朴实、待客豪爽,特别讲究待客之道。

饮食文化

佳句点睛 Punchlines

1. The soup is emphasized as the source of fresh food in Shandong cuisine.

鲁菜讲究"汤为百鲜之源"。

2. The cooking methods of quick stir-frying and braising in Shandong cuisine are praised by the world.

鲁菜中的爆、扒烹调法为世人所称道。

3. Shandong cuisine values the high quality of raw materials and **highlights**④ the original taste in seasoning.

鲁菜讲究原料质地优良,调味突出本味。

情景对话 Situational Dialogue

A: Hello, I just came to China, so I wanna try authentic Chinese food.

B: We are mainly Shandong cuisine here. I don't know what taste you like.

A: I don't know anything about Chinese food. Can you give me some advice?

B: Shandong cuisine is one of the eight major dishes in China, the taste is mainly salty and fresh and pays attention to the original flavor of the food.

A: Could you tell me more about Shandong cuisine?

B: Shandong cuisine **originated**⑤ in Shandong during the Spring & Autumn Period and the Warring States Period, which has a long history and involves many dishes.

A: That sounds very good. Could you **recommend**⑥ some representative dishes of Shandong cuisine for me?

B: Of course. I think Braised **Intestines**⑦ in Brown Sauce, Sweet and Sour Carp and Braised Sea Cucumber with **Scallion**⑧ all taste good.

A: I prefer Sweet and Sour Carp. Thank you.

B: My pleasure. I'm glad to serve you.

A：你好，我刚到中国，所以想尝尝正宗的中国菜。

B：我们这里主要是鲁菜，不知道你喜欢什么口味。

A：我对中国菜一无所知。你能给我一些建议吗？

B：鲁菜是中国八大菜系之一，味道以咸鲜为主，注重食物本味。

A：可以多为我介绍一些有关鲁菜的知识吗？

B：鲁菜起源于春秋战国时期的山东，其历史悠久，菜品众多。

A：听起来很不错，可以推荐一些鲁菜的招牌菜吗？

B：当然可以，我觉得九转大肠、糖醋鲤鱼和葱烧海参的味道都不错。

A：那就来一份糖醋鲤鱼吧。谢谢你。

B：不客气，很高兴为你服务。

 生词注解 Notes

① maintain /meɪnˈteɪn/ *vt.* 维护;保持

② seafood /ˈsiːfuːd/ *n.* 海鲜;海产食品

③ good-hearted /ˌɡʊd ˈhɑːtɪd/ *adj.* 好心肠的;仁慈的

④ highlight /ˈhaɪlaɪt/ *vt.* 使……突出;强调

⑤ originate /əˈrɪdʒɪneɪt/ *vi.* 起源

⑥ recommend /ˌrekəˈmend/ *vt.* 推荐;介绍

⑦ intestines /ɪnˈtestɪnz/ *n.* 肠;内脏

⑧ scallion /ˈskælɪən/ *n.* 青葱;冬葱

粤菜

Guangdong Cuisine

导入语 Lead-in

粤菜源自中原,又叫"广东菜",是中国八大菜系之一。粤菜由广州菜(广府菜)、潮州菜(潮汕菜)、东江菜(客家菜)三种地方菜构成,而这三种地方菜系又各具特色。广州菜擅长以山珍美味为原料,烹调成形态各异的美味佳肴;潮州菜主要以海味、河鲜和畜禽为原料,擅烹以蔬果为原料的素菜;东江菜又称客家菜,用料以肉类为主,讲究酥、软、香、浓。粤菜在历史发展中吸收了中国北方菜和西餐的技巧,口味以"清、鲜、嫩、滑、爽、香、脆"为主,可谓"中西合璧"。粤菜的招牌菜有白切鸡、红烧乳鸽、上汤焗龙虾、脆皮烧肉、清蒸东星斑、阿一鲍鱼、蒜香骨、广东早茶、煲仔饭和蜜汁叉烧等。

饮食文化

  文化剪影 Cultural Outline

Guangdong cuisine has been further developed with the southward **migration**① of some people from the Central **Plains**②, bringing with them some advanced skills. At the same time, it is located in **coastal**③ areas of Guangdong and has close exchanges with foreign countries.

随着部分中原人南迁带来了先进的技艺,粤菜也得到了进一步发展,同时广东位于沿海地区,对外交流密切。

Guangdong is located in the **subtropical**④ region, rich in products. The development of Guangdong cuisine **spans**⑤ a long period of time, with both traditional Chinese food flavors and western cooking techniques.

广东位于亚热带地区,物产丰富。粤菜的发展时间跨度大,既有中国传统的饮食风味,又有西方外来的烹饪技巧。

The main characteristics of Guangdong cuisine pay more attention to the original taste and quality of food, which can change with the seasons, mainly **manifested**⑥ in the light flavor of summer and autumn and the rich flavor of winter and spring.

粤菜的主要特色是较为注重食品的原味和质量,菜品随季节时令的变化而变化,主要表现为夏秋多清淡风味、冬春多浓郁风味。

佳句点睛 Punchlines

1. Guangdong cuisine is best at stir-frying, focusing on the control of heat and oil temperature.

粤菜最擅长小炒,注重对火候和油温的把握。

2. Guangdong cuisine can be made according to the **preference**⑦ of diners.

粤菜可以根据食客的喜好制定菜品口味。

3. Guangdong cuisine has become the most representative and **influential**⑧ dish in China in the process of continuous development.

粤菜在不断发展的过程中成为国内颇具代表性和具有世界影响的菜式。

 情景对话 Situational Dialogue

A: I've heard you went on a trip to Guangdong this summer vacation. Is there anything you can share with me?

B: The trip was really interesting, and what surprised me most was Guangdong cuisine.

A: Why Guangdong cuisine?

B: Because I find it is similar to some cuisine of the Central

Plains, and meanwhile contains the characteristics of the western cuisine.

A: The reason why Guangdong cuisine has the characteristics of the Western cuisine is that its **geographical**⑨ location is similar, so why is it similar to the cuisine of the Central Plains?

B: I heard from the local boss, because of some historical reasons, some people from the Central Plains moved south to Guangdong here, and also brought the dishes of the Central Plains.

A: It turns out that Guangdong cuisine has such a long history!

B: Guangdong cuisine is one of the eight traditional Chinese cuisines.

A: We can study these eight traditional cuisines if we have enough time.

B: Of course!

A：我听说你这个暑假去广东旅游了，有什么可以分享的吗？

B：这次旅行确实很有趣，最让我惊喜的是粤菜。

A：为什么是粤菜呢？

B：我发现它与一些中原菜式相似，同时又包含西式菜品的特点。

A：粤菜具有西式菜品特点的原因是地理位置相近，为什么会与中原菜式相近呢？

B：我听当地的老板介绍，古时候因为一些历史原因部分中原人南迁到了广东，也带去了中原的菜肴。

A: 原来粤菜的历史这么悠久啊!

B: 粤菜可是中国传统的八大菜系之一。

A: 我们有时间可以好好研究一下这八大传统菜系。

B: 当然可以!

生词注解 Notes

① migration /maɪˈgreɪʃn/　*n.* 迁移;移民

② plain /pleɪn/　*n.* 平原;朴实无华的东西

③ coastal /ˈkəʊstl/　*adj.* 沿海的;海岸的

④ subtropical /ˌsʌbˈtrɒpɪkl/　*adj.* 亚热带的

⑤ span /spæn/　*vt.* 跨越;持续

⑥ manifest /ˈmænɪfest/　*vt.* 表现;显现

⑦ preference /ˈprefrəns/　*n.* 偏爱;倾向

⑧ influential /ˌɪnfluˈenʃl/　*adj.* 有影响的;有势力的

⑨ geographical /ˌdʒiːəˈgræfɪkl/　*adj.* 地理的;地理学的

湘菜

Hunan Cuisine

导入语 Lead-in

湘菜又称"湖南菜"，是中国传统饮食文化和湖南当地饮食文化的结晶。湘菜早在汉朝时期便已形成菜系，是中国历史悠久的八大菜系之一，也是中国历史最悠久的地方菜，以湘江流域、洞庭湖区和湘西山区三种地方风味为主。湘菜制作精细，用料广，口味重，品种多，油重色浓，讲究实惠，注重香辣、香鲜、软嫩，以煨、炖、腊、蒸、炒见长。湘菜的招牌菜有剁椒鱼头、辣椒炒肉、组庵鱼翅、湘西外婆菜、金鱼戏莲、组庵豆腐、牛肉粉、衡阳鱼粉、永州血鸭、姊妹团子和安东鸡等。

文化剪影 Cultural Outline

The making method of Hunan cuisine is more **complicated**①, and there're several types of techniques in each category. Relatively speaking, Hunan cuisine is better at **simmering**②, almost reaching to the point of perfection.

湘菜制作工艺较为繁杂,每类技法多则几十种。相对来说,湘菜煨的功夫更胜一筹,几乎达到炉火纯青的地步。

Hunan cuisine consists of three local flavor schools—Xiangjiang River **Basin**③, Dongting Lake Region and Western Hunan Mountain Area. Different regions focus on different points, but the common characteristic is the spicy taste.

湘菜由湘江流域、洞庭湖区和湘西山区三种地方风味流派组成,不同的地域侧重不同,但都具有辣的特点。

Hunan cuisine has always attached great importance to the matching of raw materials and mutual **penetration**④ of taste, however the spicy taste is valued. Due to the mild and **humid**⑤ climate of Hunan Province, the people there mostly love to eat red pepper to refresh themselves and remove dampness.

湘菜历来重视原料互相搭配,滋味相互渗透,调味尤重香辣。因湖南气候温和湿润,故人们多喜食辣椒,用以提神去湿。

佳句点睛 Punchlines

1. There are many varieties of Hunan cuisine and various cooking techniques.

湘菜品种繁多,烹饪技法多样。

2. Hunan cuisine has **economical**⑥ popular dishes as well as **elegant**⑦ feast dishes.

湘菜既有经济实惠的大众菜式,也有格调高雅的宴会菜式。

3. Stir-frying is a specialty of Hunan cuisine.

爆炒是湖南人做菜的拿手好戏。

 情景对话 Situational Dialogue

A: What did you bring today?

B: I brought Mao's Braised Pork.

A: I heard that is Chairman Mao's favorite dish.

B: Yes, it is. A specialty in Hunan cuisine.

A: What is the difference with general braised pork?

B: It tastes sweet, salty and a little spicy, but not **greasy**⑧. You can have a taste of it.

A: It's so delicious, different from the braised pork I used to eat.

B: Yes, it's sweet and a little spicy.

A: Hunan cuisine is famous for its spicy flavor, isn't it?

B: It's really famous for its spiciness, but it is also very delicious.

A: But isn't it bad for your health if you eat too much of spicy food?

B: Of course not, the **characteristics**⑧ of Hunan cuisine are related to the local climate.

A: 今天你带来什么好吃的啊?

B: 我带的是毛氏红烧肉。

A: 我听说那是毛主席最爱的一道菜。

B: 是的,这是湘菜里的一道特色菜。

A: 它跟一般红烧肉有什么区别呢?

B: 毛氏红烧肉味道甜中带咸、咸中有辣、甜而不腻。你可以尝尝。

A: 太好吃了,跟我以前吃的红烧肉味道不一样。

B: 是的,有甜味也有些许辣味。

A: 湘菜著名的特色是辣,对吗?

B: 它确实以辣而闻名,但更多的是美味。

A: 你们经常吃那么辣的东西不会对身体不好吧?

B: 当然不会,湖南菜的特色与当地的气候有关。

生词注解　Notes

① complicated /ˈkɒmplɪkeɪtɪd/　*adj.* 难懂的；复杂的

② simmer /ˈsɪmə/　*vt.* 用文火炖；煨

③ basin /ˈbeɪsn/　*n.* 流域；盆地

④ penetration /ˌpenəˈtreɪʃn/　*n.* 穿透；渗透

⑤ humid /ˈhjuːmɪd/　*adj.* 湿润的；多湿气的

⑥ economical /ˌiːkəˈnɒmɪkl/　*adj.* 节约的；合算的

⑦ elegant /ˈelɪɡənt/　*adj.* 高雅的；优雅的

⑧ greasy /ˈɡriːsɪ/　*adj.* 油腻的；含脂肪多的

⑨ characteristic /ˌkærəktəˈrɪstɪk/　*n.* 特征；特色

第二部分 八大菜系

苏菜

Jiangsu Cuisine

导入语 Lead-in

苏菜又称为"淮扬菜""江苏菜",是中国八大菜系之一,主要由南京菜、淮扬菜、苏锡菜和徐海菜等地方菜组成。江苏地处中国东部温带,气候温和,地理条件优越,丰富的烹饪原料为江苏烹饪的发展提供了良好的物质基础。苏菜擅烹制鲜活淡水产品,讲究刀工,注重火候。明清时期,苏菜"南北沿运河、东西沿长江"的发展更为迅速。沿海的地理优势扩大了苏菜在海内外的影响。江苏人口密集,经济发展和贸易往来更是为烹饪事业的发展和崛起带来了巨大的推动力,苏菜的烹饪技艺也因此得以长足发展。苏菜的招牌菜有金陵烤鸭、彭城鱼丸、老鸭汤、炖生敲、霸王别姬、盐水鸭、金香饼、凤尾虾等。

 文化剪影 Cultural Outline

Jiangsu is a treasure house for arts and crafts, where there're classical gardens, Kun Opera and embroidery. Besides living, **entertainment**[①] and clothing, Jiangsu is also a paradise for **gourmets**[②].

江苏是工艺美术的宝库,那里有古典园林、昆曲和刺绣。除了生活、娱乐、服饰之外,江苏还是美食天堂。

Jiangsu cuisine has a long history and splendid culture. Jiangsu cuisine preserves original flavors and maintains freshness along with **versatility**[③]. Dish styles are exquisite and elegant, combining a fine appearance with good quality.

苏菜历史悠久,文化灿烂。苏菜保留了原有的风味,注重保持菜品新鲜和多样性。菜式精致典雅,外观精美,质量上乘。

Jiangsu cuisine is renowned for its geographic features and numerous historical **allusions**[④], which takes advantage of a rich source of fish and rice in the Yangtze River, absorbing the culinary art in northern part of China and **integrating**[⑤] the scenery of the historical sites into one.

苏菜以地理特色和众多历史典故而闻名,利用江南鱼米之乡物产丰富的优势,吸收北方烹饪技艺,集名胜古迹的文采风貌于一体。

佳句点睛 Punchlines

1. Jiangsu cuisine is elegant and beautiful in form and quality.

苏菜风格典雅,形质均美。

2. The cutting skill of Jiangsu cuisine is exquisite and the cooking methods are varied—braising, stewing, simmering, and warming.

苏菜刀工精细,烹调方法多样——炖、焖、煨、焐并举。

3. The culinary arts of Jiangsu cuisine embody the **craftsmanship**⑥ spirit, making dishes as perfect as possible.

苏菜的烹饪艺术体现了工匠精神,把菜做得尽善尽美。

情景对话 Situational Dialogue

A: Welcome to listen to our program—*This Is Suzhou*!

B: What are you gonna introduce to us today?

A: Today we're gonna talk about Jiangsu cuisine. The cooking process of Jiangsu cuisine has recently been listed in the fourth recommended list for provincial-level intangible cultural heritage, once again raising people's attention to the protection of this **culinary**⑦ school.

B: What are the characteristics of Jiangsu cuisine?

A: After hundreds of years' culinary development, Jiangsu cuisine has developed a unique flavor and fame. The emphasis of the course is on the use of high-quality ingredients, tender meat, fresh vegetables and care in preparation.

B: What are the representatives of Jiangsu cuisine?

A: Sweet and Sour Mandarin Fish, one of the representatives of Jiangsu cuisine, which is characterized by its lack of bones and the softness of its meat.

B: What's the difference between Jiangsu cuisine and other similar dishes?

A: Jiangsu cuisine is light in taste and sweeter than neighboring Yangzhou's Huaiyang cuisine. The emphasis is on the use of quality ingredients and attention to detail, meaning the dishes are all unique in color, **aroma**⑧, shape and taste.

B: Suzhou is also famous for its desserts. Such as Suzhou traditional festival dessert Steamed **Osmanthus**⑨ Cakes.

A: Yes, it's a traditional dessert during the Spring Festival in Suzhou. The cakes are produced after thirteen steps, including mixing flour, pressing, rubbing, cutting and so on.

B: There're so many delicacies in Suzhou. Welcome to Suzhou!

A: 欢迎各位收听我们今天的节目《这就是苏州》!

B: 今天你要给大家介绍什么呢?

A: 今天我们要讲苏州的美食。苏菜的烹饪过程最近被列入省级非物质文化遗产第四推荐名录,再次引起了人们对苏菜烹饪技术保护的关注。

B: 苏菜都有什么特点呢?

A: 苏菜经过数百年的发展,形成了独特的风味,享有盛誉。制作的重点是选用优质原料,肉质鲜嫩,蔬菜新鲜和准备精良。

B: 苏菜都有哪些代表?

A: 松鼠鳜鱼是苏菜的招牌菜之一,特点是没有鱼刺,肉质柔软。

B: 那苏菜跟其他同类菜有什么区别呢?

A: 苏州菜味道清淡,比邻近的扬州淮扬菜甜。使用优质配料,注重细节,每道菜肴在颜色、香气、形状和味道上都是独一无二的。

B: 苏州的甜点也很有名啊!如苏州传统节日时的甜品清蒸桂花糕。

A: 是啊!桂花蒸年糕是苏州春节期间的传统甜点。要经过十三道制作工序,包括混合面粉、压制、揉搓和切割等。

B: 苏州美食真是太多了。欢迎到苏州来!

生词注解 Notes

① entertainment /ˌentəˈteɪnmənt/ *n.* 娱乐;消遣

② gourmet /ˈɡʊəmeɪ/ *n.* 美食家;食客

③ versatility /ˌvɜːsəˈtɪlətɪ/ *n.* 多功能性;多才多艺

④ allusion /əˈluːʒn/ *n.* 典故;暗示

⑤ integrate /ˈɪntɪɡreɪt/ *vt.* 使……一体化;使……成整体

⑥ craftsmanship /ˈkrɑːftsmənʃɪp/ *n.* 技艺;工艺

⑦ culinary /ˈkʌlɪnərɪ/ *adj.* 烹饪的;烹饪用的

⑧ aroma /əˈrəʊmə/ *n.* 芳香;香味

⑨ osmanthus /ɒzˈmænθəs/ *n.* 桂花

浙菜

Zhejiang Cuisine

导入语 Lead-in

浙菜主要由以杭州、宁波、绍兴为代表的三个地方菜流派组成。浙江凭借四季时鲜不断、鱼虾海味资源丰富的优势,为浙菜的形成与发展提供了得天独厚的优越条件。

杭州菜制作精细,清秀隽美,擅长爆、炒、烩、炸等烹调技法,具有清鲜、爽嫩、精致、醇厚等特色;宁波菜善制海鲜,技法以炖、烤、蒸著称,口味鲜咸适度,菜品讲究鲜嫩爽滑,注重本味,用鱼干制品烹调更有独到之处;绍兴菜香酥绵糯、汤浓味醇,具有水乡古城之淳朴风格。浙菜的招牌菜有西湖醋鱼、干炸响铃、荷叶粉蒸肉、西湖莼菜汤、龙井虾仁、冰糖甲鱼、干菜焖肉、清汤越鸡和宋嫂鱼羹等。

文化剪影 Cultural Outline

Zhejiang Province **adjoins**① the sea in the east with **abundant**② sea products; the north is the plains; and the mid-province basin abounds with vegetables and rice. It is also the main producing area of rice and silk, known as "The Hometown of Fish and Rice".

浙江省东濒大海,盛产海味,北部是平原,中部为盆地,盛产蔬菜和稻米。浙江省又是大米与桑蚕的主要产地,素有"鱼米之乡"之称。

Chinese cuisine is characterized by diversified cooking materials and methods. Therefore, the local Jinhua Ham and Dragon Well Green Tea are both necessary **superior**③ ingredients of Zhejiang Cuisine.

中国菜的特色在于丰富的食材和独特的烹饪技巧。因此当地的金华火腿、西湖龙井茶都是浙菜烹饪中不可缺少的上乘原料。

In the pursuit of food culture, Zhejiang cuisine also absorbed the health and other elements to promote the **segmentation**④ and upgrading, driving the **expansion**⑤ of consumption.

浙菜在追求美食文化的同时,也吸纳整合了养生等元素,促进了浙菜的细分和升级,带动了消费的扩张。

饮食文化

 ## 佳句点睛　Punchlines

1. Zhejiang cuisine takes the flavor as its core and health **preservation**⑥ as its purpose.

浙菜以味道为核心，以养生为目的。

2. As an important element of Chinese culture, cuisine culture is a link to promote exchanges among people from different countries.

作为中华文化的重要元素，饮食文化是促进各国人民交流的一条纽带。

3. Chinese food is also **conveying**⑦ the Chinese values.

中餐传递着中国人的价值观。

 ## 情景对话　Situational Dialogue

A: Oh, I'm starving. I'd like to try some real Chinese cuisines. What specialities would you recommend, waiter?

B: Well, it depends. You see, there're eight famous Chinese cuisines, for example, Sichuan cuisine and Hunan cuisine.

A: They are both spicy.

B: That's right. If you like spicy dishes, you can try some.

A: They might be too spicy for me.

B: Well, there're Zhejiang cuisine and Guangdong cuisine. They are not very spicy.

A: I have heard Dongpo Pork of Zhejiang cuisine. I'll order it.

B: OK, this is your dish.

A: Waiter, could you bring us our check?

B: Here is your bill. How do you like your dishes today?

A: The soup is **tasty**[8], but the meat is a little **tough**[9].

B: I'm sorry for that. We'll improve the dishes next time.

A: 我快饿死了,我想尝尝真正的中国菜。服务员,你有什么招牌菜推荐吗?

B: 嗯,这要看情况。你知道,中国有八大菜系,比如川菜和湘菜。

A: 它们都很辣啊。

B: 是的。如果你喜欢吃辣的,你可以尝尝。

A: 对我来说可能太辣了。

B: 我们也有浙菜和粤菜,它们不是很辣。

A: 我听说过浙菜的东坡肉,我要点这个。

B: 好的,这是你点的菜。

A: 服务员,买单。

B: 这是你的账单。你觉得今天的饭菜怎么样?

A: 汤很好喝,但肉有点儿老。

B: 很抱歉。我们下次会改进的。

 生词注解 Notes

① adjoin /əˈdʒɔɪn/ *vt.* 毗连；邻接

② abundant /əˈbʌndənt/ *adj.* 丰富的；充裕的

③ superior /suːˈpɪərɪə(r)/ *adj.* 上好的；优秀的

④ segmentation /ˌsegmenˈteɪʃn/ *n.* 分割；分裂

⑤ expansion /ɪkˈspænʃn/ *n.* 阐述；扩张物

⑥ preservation /ˌprezəˈveɪʃn/ *n.* 保存；保留

⑦ convey /kənˈveɪ/ *vt.* 传达（信息）；表达（思想）

⑧ tasty /ˈteɪstɪ/ *adj.* 美味的；高雅的

⑨ tough /tʌf/ *adj.* （肉）嚼不动的；强硬的

闽菜

Fujian Cuisine

导入语 Lead-in

闽菜是中国八大菜系之一，源于福建省。闽菜淡雅、鲜嫩、醇香、隽永，保持原味、鲜味的做法是其他菜式无法比拟的。福建的经济文化在南宋以后逐渐发展起来，清中叶后闽菜逐渐为世人所知。闽菜由福州菜、闽南菜和闽西菜三部分组成。福州菜流行于闽东、闽中和闽北地区；闽南菜广传于厦门、泉州、漳州和闽南金三角；闽西菜则盛行于闽西客家地区，极富乡土气息。闽菜口味偏重甜、酸和清淡，常用红糟调味。闽菜的招牌菜有佛跳墙、鸡汤氽海蚌、八宝红鲟饭、太极芋泥、醉排骨、龙身凤尾虾、荔枝肉、五柳居等。

 文化剪影　Cultural Outline

"Min" is an **abbreviation**① for Fujian. Fujian cuisine is cuisine which originated in Minhou County, Fujian Province. It is represented by the cuisines of Fuzhou, Quanzhou, Xiamen and other places, one of the major cuisines of China, enjoying the pride of place among the culinary cultural treasures of China.

"闽"是福建的简称,闽菜起源于福建省闽侯县。它以福州、泉州、厦门等地的菜肴为代表,是中国烹饪的主要菜系之一,在中国烹饪文化宝库中占有重要一席。

The soup is a very important element of Fujian cuisine, which is **associated**② with the cooking of sea food and traditional eating customs. For a long time, chefs of Fujian cuisine have attached great importance to the ensuring of fresh quality, pure flavors and **nourishment**③ of raw ingredients.

汤菜是闽菜中非常重要的元素,与其多烹制海鲜和传统食俗有关。长期以来,闽厨非常重视菜品的质鲜、味纯和滋补。

Fujian is **advantageously**④ situated near both mountains and coastal areas. There're mountainous areas to the north of Fujian and its coastal area is in the south. Therefore, the major ingredients for cooking are sea food and food from **mountainous**⑤ regions.

福建地理条件得天独厚,依山傍海,北部多山,南部面海,因此烹饪原料以海鲜和山珍为主。

佳句点睛 Punchlines

1. Fujian cuisine's best cooking techniques are stir-frying, sautéing, frying and simmering.

闽菜的烹调法擅长炒、溜、煎和煨。

2. Fujian cuisine's cooking process is exquisite and its seasoning is of great importance.

闽菜烹调细腻,特别注重调味。

3. Fujian cuisine is mainly about seafood with sweet, sour, salty and tasty flavors, it has beautiful color and yummy taste.

闽菜以海鲜为主,口味以甜、酸、咸、香为主,色美味鲜。

情景对话 Situational Dialogue

A: There's a popular summary of Chinese cuisine of "being sweet in the south, salty in the north, sour in the west and spicy in the east." Do you know it?

B: Yes, the local dish varies from region to region because China is a vast country with **diverse**① climates and customs.

A: Well, as for Chinese cuisine, there're five features of it—

namely, color, flavor, taste, shape, and poetry.

B: I agree with you.

A: My favorite food is Fotiaoqiang (Steamed **Abalone**⑦ with Shark Fin and Fish Maw in Broth). Do you know which cuisine it belongs to?

B: Yes, it belongs to Fujian cuisine.

A: Fujian cuisine has won a reputation for its freshness, tenderness, softness and smoothness with **mellow**⑧ fragrance.

B: Fujian cuisine is distinguished for its choice of seafood. The most distinct feature is the "steaming" technique.

A: Have you ever had Fujian cuisine before?

B: To be honest, I haven't yet.

A: Let's go to the restaurant to enjoy it together.

B: OK, let's go.

A: 对中国菜有这样一句总结:"甜在南方、咸在北方、酸在西方、辣在东方。"你知道吗?

B: 知道,因为中国幅员辽阔,各地气候和风俗不同,口味也不同。

A: 中国菜有五个特点,就是色、香、味、形和意。

B: 我同意。

A: 我最喜欢的一道菜是佛跳墙。你知道它属于哪个菜系吗?

B: 我知道,这是闽菜。

A: 闽菜以鲜嫩、柔滑、醇香而著称。

B：闽菜以海鲜闻名，最明显的特征是"糟"的特色。

A：你以前吃过闽菜吗？

B：说实话，还没有。

A：咱们一起去餐厅吃吧。

B：好的，咱们走吧。

 生词注解 Notes

① abbreviation /əˌbriːvɪˈeɪʃn/ n. 缩写；缩写词

② associated /əˈsəʊsɪeɪtɪd/ adj. 关联的；联合的

③ nourishment /ˈnʌrɪʃmənt/ n. 食物；营养品

④ advantageously /ˌædvənˈteɪdʒəslɪ/ adv. 有利地；方便地

⑤ mountainous /ˈmaʊntənəs/ adj. 多山的；巨大的

⑥ diverse /daɪˈvɜːs/ adj. 不同的；相异的

⑦ abalone /ˌæbəˈləʊnɪ/ n. 鲍鱼

⑧ mellow /ˈmeləʊ/ adj. 芳醇的；圆润的

饮食文化

徽菜

Anhui Cuisine

导入语 Lead-in

徽菜又称为"徽帮菜"或"徽州风味"。徽菜始于南宋时期的徽州府,因徽州而得名,随着徽商的快速发展而闻名遐迩,促使徽菜馆遍布全国各地。徽州地处山区,蕴藏丰富的物产资源,质地优良的原料为徽菜的形成与发展提供了良好的物质基础。同时,特有的地理位置和人文气韵使徽菜更具地方特色。经过历代名厨的创新与发展,如今徽菜已经集中了安徽各地的风味特色,逐步形成了独具一格、自成一体的著名菜系。徽菜的招牌菜有徽州毛豆腐(虎皮毛豆腐)、红烧臭鳜鱼、火腿炖甲鱼、徽州一品锅、方腊鱼、问政山笋、蜜汁山芋、黄山炖鸽、中和汤等。

文化剪影 Cultural Outline

The establishment of Anhui cuisine is closely related to the unique geographical position, humanistic background and eating customs of ancient Huizhou, and **flourished**① because of the development of Huizhou merchants.

徽菜的形成与江南古徽州特有的地理位置、人文背景、饮食习俗紧密相连,又因徽商的发展而兴盛。

In the process of long-term development, Anhui cuisine has gradually **accumulated**② a set of cooking techniques, especially the duration and degree of heating, which is highly distinctive. Therefore, Anhui cuisine has also become well-known all over the world with the increasing development of Huizhou **merchants**③.

徽菜在长期发展过程中逐渐积累了一整套烹调技法,尤其是厨师对火候的掌握,极具特色。因此,在徽商日渐发展的同时,徽菜也闻名天下。

The main features of Anhui cuisine are versed in frying, stewing, steaming, stressing on the original taste, heating degree, culture and health of the cuisine. Moreover, Anhui cuisine often **integrates**④ good meanings into it so that the dishes are full of cultural **foundation**⑤.

徽菜的主要特色为擅长烧、炖、蒸,重本味、重火功、重文化、讲食补。徽菜常常把美好寓意融入其中,使菜品充满文化底蕴。

 佳句点睛　Punchlines

1. Anhui cuisine has always been a high pursuit of taste.
徽菜历来就对口味有很高的追求。

2. Anhui cuisine pays great attention to the natural characteristics of cooking materials.
徽菜十分注重烹饪原料的自然特色。

3. A major feature of Anhui cuisine is ensuring **freshness**⑥.
保证鲜活是徽菜的一大特点。

 情景对话　Situational Dialogue

A: Could you tell me something about your hometown cuisine?

B: I come from Anhui. A Chinese saying goes that "food is the first necessity of people". I'd like to introduce you some knowledge of Anhui cuisine if you don't mind.

A: I've heard some, but I haven't learned in details. Do they have any special features?

B: Anhui cuisine is one of the eight major cuisines in China, whose main features are the cooking techniques, including cutting,

duration[7] and operating skill.

A: That sounds very **attractive**[8] to me. What is your favorite dishes?

B: My favorite is Stewed Soft-Shell Turtle with Ham.

A: What a **distinctive**[9] name!

B: The name of Anhui cuisine often has some meaning of blessing. In addition, Huizhou is rich in rare animals and vegetables while local materials can also ensure the freshness of the food.

A: I'm glad to know that. Thank you for your introduction.

B: My pleasure.

A：你可以介绍一下你的家乡菜吗？

B：我来自安徽。中国有句老话叫"民以食为天"。如果你不介意，我想先介绍有关徽菜的一些知识。

A：我听说过一些，但没有详细了解过。徽菜都有什么特色？

B：徽菜是中国八大菜系之一。它的主要特色是烹饪技法，包括刀工、火候和操作技术。

A：听起来很吸引人，你最喜欢的一道徽菜是什么呢？

B：我最喜欢的是清炖马蹄鳖。

A：多么有特色的名字啊！

B：徽菜的菜名常带一些祝福的含义。除此以外，徽地盛产这些山珍野味，就地取材还可以保证食材鲜活。

A：很高兴知道这些，谢谢你的介绍。

B：不客气。

饮食文化

 生词注解 Notes

① flourish /ˈflʌrɪʃ/ vt. 繁荣；昌盛

② accumulate /əˈkjuːmjəleɪt/ vt. 积累；(数量)逐渐增加

③ merchant /ˈmɜːtʃənts/ n. 商人

④ integrate /ˈɪntɪɡreɪt/ v. 使……结合；使……一体化

⑤ foundation /faʊnˈdeɪʃn/ n. 基础；地基

⑥ freshness /ˈfreʃnəs/ n. 新鲜；精神饱满

⑦ duration /djuˈreɪʃn/ n. 持续的时间

⑧ attractive /əˈtræktɪv/ adj. 诱人的；有魅力的

⑨ distinctive /dɪˈstɪŋktɪv/ adj. 独特的；有特色的

第三部分　特色美食

Part Ⅲ　Special Cuisines

北京烤鸭

Beijing Roast Duck

 导入语 Lead-in

北京烤鸭被誉为中国菜肴的"国粹",具有深厚的历史渊源和人文底蕴。北京烤鸭是起源于中国南北朝时期的宫廷食品,明朝嘉靖年间传到民间。烤鸭之所以美味,是因其原料为品种名贵的北京鸭。北京烤鸭因不同的烤制方式而分为"挂炉烤鸭"和"焖炉烤鸭",其用料优质,果木炭火烤制,色泽红润,肉质肥而不腻,外脆里嫩。享用北京烤鸭,为人们营造出一种享用御膳般的心理满足。

 文化剪影 Cultural Outline

Beijing Roast Duck **evolved**① from the court food in the Southern and Northern dynasties. For thousands of years, its roasting process has been continuously improved. With the more modern production process, its flavor is getting more delicious and suitable for people's tastes and **consumption**② ideas.

北京烤鸭由南北朝的宫廷食品演变而来。千百年来,北京烤鸭的烤制过程也在不断改进,生产过程更加现代化,风味更加鲜美,更加适合现代人的口味和消费观念。

As the representative of Beijing cuisine, Beijing Roast Duck is characterized by bright red skin, tender meat, mellow taste, and fatness but with no **greasiness**③, fully displaying the wisdom of the Chinese people's earliest "roasting" method in cuisine.

北京烤鸭作为北京菜的代表,其色泽红艳、肉质细嫩、味道醇厚、肥而不腻,充分展示出中国人最早掌握的烹饪法——"烤"的智慧。

Unique roasting techniques, exquisite **ingredients**④, on-site knife skills and **slap-up**⑤ eating methods make the taste of Beijing Roast Duck a kind of unique artistic enjoyment.

独特的烤制手艺、精致的配料、现场刀工献技、考究的吃法,使品尝北京烤鸭成为一种独特的艺术享受。

佳句点睛 Punchlines

1. Beijing Roast Duck evolved from the court food in the Southern and Northern dynasties.

北京烤鸭由南北朝时期的宫廷食品演变而来。

2. Beijing Roast Duck is characterized by bright red skin, tender meat, mellow taste, and fatness but with no greasiness.

北京烤鸭具有色泽红艳、肉质细嫩、味道醇厚、肥而不腻的特点。

3. Tasting Beijing Roast Duck has become a kind of unique artistic enjoyment.

品尝北京烤鸭已经成为一种独特的艺术享受。

 ## 情景对话 Situational Dialogue

A: Can I help you, sir?

B: Let me see your menu. Oh, the roast duck looks pretty good. Order one first.

A: OK. It is the specialty in Quanjude.

B: Could you tell us how it is made?

A: Certainly. First, the roast duck in our restaurant is very carefully selected. After it is cleaned and dressed, it will be put in a large oven and braised. We use fruit wood as fuel, so that the duck has a unique **aroma**⑥. When the skin turns golden, it is ready to be served with scallions, cucumbers, hoisin sauce and pancakes.

B: Wow, I'm hungry for it.

A: Do you need anything more, sir?

B: Yes. Preserved eggs with peanuts, **sautéed**⑦ cabbage with mushrooms and fried shrimps.

A: OK. Tea or fruit juice?

B: A cup of black tea, please.

A：先生，您需要点什么？

B：让我看看菜单。噢，烤鸭看起来挺不错的，先点一只烤鸭。

A：好的，这是全聚德的特色菜。

B：你能否介绍一下烤鸭是如何做成的？

A：当然可以。首先，我们店烤的鸭子都是经过精挑细选的。在洗净和清理干净内脏后，将其放在一个大炉内焖烤。我们采用果木作燃料，这样烤出来的鸭子有一种独特的香味。当鸭皮呈金黄色时，就可以用京葱、黄瓜、甜面酱和薄饼同食了。

B：哇，我都迫不及待想吃了。

A：先生还要点儿什么？

B：皮蛋花生、油菜香菇和油爆河虾。

A：好的。喝茶还是喝果汁？

B：请来杯红茶。

生词注解 Notes

① evolve /ɪˈvɒlv/   *vi.* 演变；进化

② consumption /kənˈsʌmpʃn/   *n.* 消费；消耗

③ greasiness /ˈɡriːsɪnəs/   *n.* 油腻；多脂

④ ingredient /ɪnˈɡriːdɪənt/   *n.* （尤指烹饪）原料；（成功的）要素

⑤ slap-up /ˈslæpʌp/   *adj.* 讲究的；高档的

⑥ aroma /əˈrəʊmə/   *n.* 香味；芳香

⑦ sauté /ˈsəʊteɪ/   *vt.* 油煎

饮食文化

老北京炸酱面

Traditional Beijing Noodles with Soybean Paste

导入语　Lead-in

炸酱面的主要食材是面粉、猪油和干黄酱。首先将面条煮熟，放入刚汲出的凉水中，然后再将肉末豆酱炒香，加上煮熟的青豆或大豆以及几种不同类的蔬菜丝拌在一起食用。除了扑鼻的醇厚浓香外，红萝卜脆、绿黄瓜爽、黄豆瓣绵、白豆芽嫩，还有卧在面条下的肉丁，让吃面变成了一种妙不可言的味蕾享受。如今，"炸酱面""大碗茶""烤鸭"已经共同构成了老北京饮食习俗的代名词，也是北京老百姓饮食文化中的标志性符号。

文化剪影 Cultural Outline

The essence of Traditional Beijing Noodles with Soybean Paste lies in the flavor of the sauce and its **tenacity**① and **elasticity**②. The green bean sprouts, cucumber shreds and turnip shreds should be blended with the noodles, which will surely impress you.

老北京炸酱面讲究的是酱的味道和面的筋道。将炒好的酱与新鲜的绿豆芽、黄瓜丝、萝卜丝等一起拌着吃,一定会让你回味无穷。

A variety of vegetable shreds made up for the lack of vitamin in fried sauce, which can also control the eating speed, avoid swallowing too quickly and aid **digestion**③ and absorption.

多种蔬菜丝既可以弥补炸酱缺乏维生素的不足,又可以有效控制吃面的速度,避免过快吞咽,有助于消化吸收。

The history of Traditional Beijing Noodles with Soybean Paste is only about one hundred years, but it reflects the history of an era, embodies a kind of food interest and **consigns**④ many sincere feelings for old Beijingers.

老北京炸酱面的历史虽然只有一百年左右,但对老北京人而言,它反映了一个时代的历史,体现了一种食趣,寄托了诸多情愫。

饮食文化

 佳句点睛 Punchlines

1. As a sign of the time-honored Beijing cuisine culture, Traditional Beijing Noodles with Soybean Paste is very popular among the people in Beijing.

作为老北京饮食文化的一个招牌,炸酱面深受北京百姓的喜爱。

2. Traditional Beijing Noodles with Soybean Paste has a different style of eating from all other kinds of noodles.

老北京炸酱面具有与其他面条不同的食法风格。

3. It seems that Traditional Beijing Noodles with Soybean Paste are closely related to the theory of health preservation of traditional Chinese medicine.

老北京炸酱面似乎还跟中医养生理论有着千丝万缕的联系。

 情景对话 Situational Dialogue

A: How are you? I haven't seen you for a long time!

B: I'm fine. How are you doing?

A: Pretty well. Let's get together for lunch. I know that there's a **fantastic**⑤ restaurant outside the North Gate.

B: Which restaurant?

A: Xiaohuajiao.

B: Oh, what a pity. I wish I could, but spicy food is not fit to my appetite. I'd like to have some snacks.

A: Well, how about Traditional Beijing Noodles with Soybean Paste? It's a traditional Beijing snack.

B: How does it taste?

A: It's very **yummy**[6] and appetizing. The bean sauce has to be a mixture of dry yellow paste and sweet bean paste. The **lean**[7] and fat pork has to be in the proportion of three to seven, which is then diced into tiny pieces. The fat meat goes in the pan first until the grease is released. It is then followed by the diced lean pork, with **minced**[8] ginger and scallion. Keep stirring until the mixed aroma of meat, scallion and ginger is set off.

B: Whoa! My mouth is watering! Let's have a try.

A: 你好吗？我好久没见到你了！

B: 我很好。你呢？

A: 很好。我们一起吃午饭吧，我知道北大门外有一家很棒的餐厅。

B: 哪家餐厅？

A: 小花椒。

B: 噢，真遗憾，我很想去，但我不能吃辣，我想吃点儿小吃。

A: 老北京炸酱面怎么样？这是北京的传统小吃。

B: 它味道如何？

A: 既好吃又开胃。它选用的豆瓣酱是干黄酱和甜豆瓣酱的混合物。瘦肉和肥肉的比例是三比七，然后切成小块。肥肉先放在锅里，直到出油。然后放入瘦肉丁、姜末和葱末。继续搅拌，直到肉、葱和姜的混合香味释放出来。

B: 哇！我都流口水了！我们去尝尝吧。

生词注解 Notes

① tenacity /təˈnæsətɪ/ *n.* 韧性；黏性

② elasticity /ˌiːlæˈstɪsətɪ/ *n.* 弹性；灵活性

③ digestion /daɪˈdʒestʃən/ *n.* 消化；领悟

④ consign /kənˈsaɪn/ *vt.* 寄存

⑤ fantastic /fænˈtæstɪk/ *adj.* 极好的；极出色的

⑥ yummy /ˈjʌmɪ/ *adj.* 好吃的；美味的

⑦ lean /liːn/ *adj.* 瘦的；脂肪少的

⑧ minced /mɪnst/ *adj.* 切碎的；切成末的

冰糖葫芦

Candied Haws on a Stick

  导入语 Lead-in

冰糖葫芦历史悠久，堪称最原始、最传统的糖果，是中国特色民俗食品之一。冰糖葫芦由山楂上裹着一层凝固的糖制成，食之酸甜可口，是北京的标志性小吃  之一。起初，每串儿上只有两个大小不一的山楂果，大个儿的在下面，小个儿的在上面，中间用一根竹签穿起，像个葫芦似的，故名"糖葫芦"。山楂富含多种营养物质、膳食纤维和活性物质，是一种养生食品。冰糖葫芦不仅富有民俗特色，许多人也认为冰糖葫芦预示着生活幸福、甜甜蜜蜜、红红火火。

文化剪影 Cultural Outline

The sugar used to make Candied Haws on a Stick should be boiled **foamed**①. Then, the stick with haws gently **rotated**② on the foam of hot sugar and was wrapped with a thin layer. Put the glutinous haws on water plate to cool down, and the **transparent**③ Candied Haws on a Stick is done.

制作冰糖葫芦的糖要熬到起泡儿,呈金黄色。糖熬好后,将穿好的山楂在热糖上轻轻转动,裹上薄薄一层,将蘸满黏稠糖水的山楂串放到水板上冷却,便制成了晶莹透明的冰糖葫芦。

Candied Haws on a Stick is a traditional Chinese snack. The red haw are lined one by one on a clean stick, liking a string of red lanterns in the sunshine. The outer layer of sugar is **crisp**④, but the haws are soft inside.

冰糖葫芦是一种中国传统小吃。红彤彤的山楂果一个个排在竹签上,在阳光下像一串红灯笼。外层的糖皮是脆的,但里面的山楂是软的。

Candied Haws on a Stick is suitable for young and old and has many medicinal effects. The sourness of haws can always give you a good **appetite**⑤, and the rich in Vitamin C and Vitamin E can make your skin **smoother**⑥.

冰糖葫芦老少皆宜，具有多种药用价值。山楂果的酸味具有开胃消食的作用，蕴含丰富的维生素C和维生素E，能让皮肤更加光滑。

佳句点睛 Punchlines

1. The key point of making Candied Haws on a Stick is the technique of boiling sugar.

制作冰糖葫芦的重点在于熬糖。

2. Candied Haws on a Stick is not only a kind of delicious food, but also the feelings of generations of old Beijingers.

冰糖葫芦不仅是一种美食，而且是一代代老北京人的情怀。

3. Candied Haws on a Stick is not only sweet and sour, but also **affordable**[7].

冰糖葫芦既酸甜可口，又经济实惠。

情景对话 Situational Dialogue

A: What's that over there? Something like the red meat ball?

B: Oh, That's Bingtang Hulu (Candied Haws on a Stick). It is also one of my favorite foods. You should try it.

A: It's sweet on the outside and a little bit sour and **juicy**[8] on the inside. Yummy!

085

B: It is common snacks in winter in North China. They are usually strung with haws. It tastes sour and sweet.

A: Bingtang Hulu? Sounds interesting!

B: It's also called Tanghulu.

A: It must be complex to make this yummy snack.

B: Yes, the haws strung on a bamboo stick were dipped in sugar, which hardens rapidly in the wind.

A: Chinese people are really smart! How can you think of such a complicated method to make it?

B: It is said that a stick of sugar-coated haws originated from the Song Dynasty, the empress wouldn't like to eat food after illness, then the emperor was very worried. At that time a doctor suggested cooking food with haw and rock sugar and ate them before the meal, and later the empress recovered after one month. The **formula**[⑨] became royal's top secret, but it spread to folks finally.

A: Oh, I see.

A: 那是什么？像红色小肉球一样？

B: 那是冰糖葫芦，也是我喜爱的食物之一。你应该尝尝。

A: 外面甜甜的，里面有酸酸的汁。太好吃了！

B: 这是中国北方冬天常见的小吃，一般用山楂串成，吃起来又酸又甜又冰。

A: 冰糖葫芦？听起来很有趣！

B: 冰糖葫芦又叫糖葫芦。

A: 做这道美味的小吃肯定很复杂吧?

B: 是的,它是将山楂用竹签串成串后蘸上麦芽糖稀,糖稀遇风迅速变硬。

A: 中国人真聪明! 你们怎么能想到用这么复杂的方法去做这个呢?

B: 冰糖葫芦起源于宋朝,据说皇后患病难以进食,皇帝心急如焚,当时,一名医生建议用山楂和冰糖熬制食物,每餐饭前食用,结果一个月后皇后就康复了。这个秘方被皇室视为最高机密,但它后来流传到了民间。

A: 噢,我明白了。

生词注解 Notes

① foamed /fəʊmd/ *adj.* 泡沫状的

② rotate /rəʊˈteɪt/ *vi.* 旋转;转动

③ transparent /trænsˈpærənt/ *adj.* 透明的;明显的

④ crisp /krɪsp/ *adj.* 脆的;易碎的

⑤ appetite /ˈæpɪtaɪt/ *n.* 食欲;嗜好

⑥ smooth /smuːð/ *adj.* 光滑的;顺畅的

⑦ affordable /əˈfɔːdəbl/ *adj.* 负担得起的

⑧ juicy /ˈdʒuːsɪ/ *adj.* 多汁的;生动有趣的

⑨ formula /ˈfɔːmjələ/ *n.* 配方

驴打滚儿

Glutinous Rice Rolls Stuffed with Red Bean Paste

 导入语　Lead-in

　　驴打滚儿是老北京深受欢迎的传统风味小吃，俗称"豆面糕"，在东北和天津地区也很流行。首先将黄米面或糯米糕蒸熟，然后抹上红豆沙卷起来，再切成小块，外滚黄豆面，撒上白糖就算制作完成了。驴打滚儿表面金黄，甘香扑鼻，断面可见黄米环绕着褐色豆馅，吃起来软糯劲道。其成品黄、白、红三色分明，煞是好看。因其制作工序中撒黄豆面时，既像北京郊外驴子撒欢打滚儿时扬起的漫天黄土，又似驴子在黄土上打滚，粘上了一层黄泥，故得名"驴打滚儿"。

文化剪影 Cultural Outline

It is said that Glutinous Rice Rolls Stuffed with Red Bean Paste was first made in the Qing Dynasty and became a kind of food for the **imperial**① court. Later, it was spread around and loved by the public. In the long history of development, its raw materials have been developing towards the taste of the public.

相传驴打滚儿最早制作于清朝,起初为宫廷美食,之后传播到民间并为大众所喜爱。在长期的历史发展进程中,其制作原料越发契合大众口味。

Glutinous Rice Rolls Stuffed with Red Bean Paste, as one of Beijing special snacks, the raw materials include yellow rice flour, soybean flour, sugar, red bean **paste**② and so on. The red bean paste should be evenly **distributed**③ in the making process.

作为北京特色小吃之一,驴打滚儿的原料有黄米面、黄豆面、白糖、豆沙等。制作时要求馅卷均匀,层次分明。

Glutinous Rice Rolls Stuffed with Red Bean Paste is a long roll made by yellow rice with red bean paste inside, looking golden in **appearance**④, characterized by appetizing, sweet and **glutinous**⑤, with a strong flavor of soybean powder.

驴打滚儿是用黄米面夹馅卷成长卷,外表呈金黄色,特色是香、甜、黏，具有浓郁的黄豆粉香味。

佳句点睛 Punchlines

1. The raw materials of Glutinous Rice Rolls Stuffed with Red Bean Paste have been developing towards to **diversification**⑥.

驴打滚儿的制作原料一直在朝着多元化方向发展。

2. Glutinous Rice Rolls Stuffed with Red Bean Paste is a traditional snack for all ages.

驴打滚儿是一款老少皆宜的传统小吃。

3. The bean paste of Glutinous Rice Rolls Stuffed with Red Bean Paste **melts**⑦ in your mouth, sweetening into your heart.

驴打滚儿的豆馅入口即化，香甜入心。

情景对话 Situational Dialogue

A: Hi, what are you looking at?

B: I'm searching for Beijing cuisine.

A: Why do you look so **confused**⑧?

B: Because there's a delicious food called Glutinous Rice Rolls Stuffed with Red Bean Paste. What's that?

A: Ha-ha, that's one of my favorite food.

B: Really? Can you tell me something about it?

A: Of course, it's actually a **dessert**⑨.

B: A dessert? What is it made of?

A: The main raw material is soybean, together with sugar and red bean paste.

B: How does it taste?

A: It tastes tender and sweet. We can taste it together when we have time.

B: Sounds good. I'm looking forward to it.

A: 嗨,你在看什么呢?

B: 我在搜索北京美食。

A: 你为什么看起来那么疑惑?

B: 因为我看到有一种美食叫"驴打滚儿",那是什么东西?

A: 哈哈,那是我最爱吃的美食之一。

B: 真的吗? 那你可以给我讲讲吗?

A: 当然可以,它其实是一种甜品。

B: 甜品? 它是用什么做的呢?

A: 主要原料为黄豆面,此外还有白糖、豆沙等。

B: 它味道怎么样?

A: 吃起来既软糯又香甜。有时间我们可以一起去品尝一下。

B: 听起来很不错,我很期待。

饮食文化

生词注解 Notes

① imperial /ɪmˈpɪərɪəl/ *adj.* 皇帝的；至高无上的

② paste /peɪst/ *n.* 面团；肉（或鱼等）酱

③ distributed /dɪˈstrɪbjuːtɪd/ *vt.* 分配；分布

④ appearance /əˈpɪərəns/ *n.* 外观；外貌

⑤ glutinous /ˈgluːtənəs/ *adj.* 黏的；胶质的

⑥ diversification /daɪˌvɜːsɪfɪˈkeɪʃn/ *n.* 多样化

⑦ melt /melt/ *vt.* 使……融化；使……熔化

⑧ confused /kənˈfjuːzd/ *adj.* 困惑的；混乱的

⑨ dessert /dɪˈzɜːt/ *n.* 餐后甜点；甜点心

狗不理包子

Go Believe

 导入语 Lead-in

清朝末年,天津有一家蒸食铺,店主名叫高贵友,从小性格倔,父母给他起个小名叫"狗子"。高贵友心灵手巧,做出的包子味道鲜美,独具特色。当时,来往于南北运河的小商小贩和平民百姓是蒸食铺的常客,人们都亲热地直呼"狗子卖包子,不理人"。久而久之,就直接叫"狗不理"包子了。俗话说"包子好吃不在褶",但狗不理包子褶花匀称,每个包子的褶花不少于十五个,均匀整齐,外形美观,看上去雪白油嫩,活像一朵朵含苞欲放的白菊花。

饮食文化

 文化剪影　Cultural Outline

Go Believe, founded in 1858 and **branded**①, is a traditional snack of the Han nationality in Tianjin as one of China's time-honored brands.

狗不理包子是天津汉族的传统风味小吃,始创于公元1858年,是中华老字号之一。

The stuffing of Go Believe has different types, such as fresh meat, seafood, braised pork with soy sauce and vegetable. The traditional handmade craft of Go Believe has been included in the *National **Intangible**② Cultural **Heritage**③ List*.

狗不理包子种类多样,有鲜肉包、海鲜包、酱肉包、素包子等。狗不理包子传统手工制作技艺被列入《国家级非物质文化遗产名录》。

Go Believe has had a history of more than 100 years, characterized by finely **selected**④ materials and exquisite ingredients, which have still **maintaining**⑤ the traditional flavor so far.

狗不理包子已有一百多年的历史,特点是选料精细,讲究配料,至今依然保持传统风味。

 佳句点睛　Punchlines

1. Go Believe has the strict quality standards in order to maintain

the traditional flavor of baozi.

为了保持包子的传统风味,狗不理包子有严格的质量标准。

2. Go Believe is famous in the world by its own unique flavor.

狗不理包子因其独特风味而闻名于世。

3. Go Believe is **economical**⑥, but rich in national characteristics.

狗不理包子经济实惠,但却富有民族特色。

 情景对话 **Situational Dialogue**

A: Hi, do you prefer Chinese food or Western food?

B: Well, to be honest with you, Chinese food is really different from Western food.

A: Are you used to the food here?

B: I'm not really used to it yet.

A: What's your favorite Chinese dish?

B: Like most foreigners, I really like Jiaozi and Baozi.

A: Have you ever tried Go Believe? It's famous in China.

B: I have tried once. It's delicious. The most important is that we can enjoy the food inside a small steamed bun.

A: Yes, the stuffing of Go Believe is rich, which is really a temptation for everyone.

B: Sometimes we need to change the diet, especially after having vegetables for quite a long time.

饮食文化

A: That's right.

B: Thank you for telling me so much knowledge.

A: 嗨,你喜欢中餐还是西餐?

B: 老实说,中餐和西餐真的不一样。

A: 你习惯这里的食物吗?

B: 我还不太习惯。

A: 你最喜欢的中国菜是什么?

B: 像大多数外国人一样,我真的很喜欢饺子和包子。

A: 你尝过狗不理包子吗?它在中国很有名。

B: 我尝过一次,很好吃。我们可以在一个小小的笼屉里享用美味。

A: 是的,狗不理包子馅料丰富,真诱人。

B: 有时我们需要调换一下饮食,尤其是在长时间吃素之后。

A: 没错。

B: 谢谢你告诉我这么多知识。

生词注解　Notes

① brand /ˈbrænd/　vt. 打烙印于……;加商标于……

② intangible /ɪnˈtændʒəbl/　adj. 无形的

③ heritage /ˈherɪtɪdʒ/　n. 遗产;继承物

④ select /sɪˈlekt/　vt. 选择;挑选

⑤ maintain /meɪnˈteɪn/　vt. 维持;继续

⑥ economical /ˌiːkəˈnɒmɪkl/　adj. 经济的;节约的

耳朵眼炸糕

Earhole Fried Cake

 导入语 Lead-in

耳朵眼炸糕起源于晚清光绪年间,是天津有名的传统风味小吃,与狗不理包子、桂发祥十八街麻花并称为"天津三绝"。小小炸糕本是不登大雅之堂的街边美食,后因其价廉物美、料精味好而传遍大街小巷,又因摊位临近耳朵眼胡同而得名,遂成为妇孺皆知的品牌,流传至今。耳朵眼炸糕素以金黄色的外形和香甜软糯的口感而广受大众喜爱,与之相关的传说也为人津津乐道,小小的炸糕中蕴含着丰富的津味文化,成为天津传统文化的物质载体。

文化剪影 Cultural Outline

Earhole Fried Cake originated in the reign of Guangxu of the late Qing **Dynasty**①. It's named Earhole Fried Cake because it is near Earhole Hutong. Since then it has been spread, has repeatedly won national awards and has been one of the famous snacks in China.

耳朵眼炸糕起源于晚清光绪年间,因其摊位临近耳朵眼胡同而得名,从此流传开来,屡获国家奖项,已经成为"中华名小吃"。

Earhole Fried Cake is a traditional snack of Tianjin, fully **reflecting**② that the Chinese people have applied the cooking technique of frying to the making of traditional cakes.

耳朵眼炸糕是天津传统特色小吃,充分体现出中国人将"炸"的烹饪技艺炉火纯青地运用到传统糕点的制作当中。

With a long history, fine and high-quality materials, **nourishing**③ and **invigorating**④ food therapy value, and **affordable**⑤ price, Earhole Fried Cake has become one of Tianjin's enduring flavor delicacies.

悠久的历史渊源,精细优质的用料,滋补益气的食疗价值,经济实惠的价格,使耳朵眼炸糕成为天津经久不衰的风味美食之一。

第三部分 特色美食

佳句点睛 Punchlines

1. Earhole Fried Cake is a famous street snack.

耳朵眼炸糕是著名的街边小吃。

2. Earhole Fried Cake, Go Believe and Guifaxiang Fried Dough Twists are knoun as the three unique snacks of Tianjin.

耳朵眼炸糕与狗不理包子、桂发祥十八街麻花并称"天津三绝"。

3. Earhole Fried Cake **adheres**⑥ to the selection of fine materials, always giving priority to quality.

耳朵眼炸糕坚持选料精细，始终把质量放在第一位。

情景对话 Situational Dialogue

A: Your snack looks so **tempting**⑦. I'll have one.

B: Just a moment. I'll fry one for you.

A: What's the name of your snack?

B: Earhole Fried Cake.

A: That's interesting. How is Earhole Fried Cake made?

B: It's very complicated. First, the glutinous rice is ground into a paste after being soaked in water, **fermentation**⑧ as dough after adding water and **alkali**⑨, then fill up with the red bean. Put it into the hot oil quickly and turn over frequently.

A: That sounds yummy.

B: It's ready. Here's your Earhole Fried Cake.

A: How much is it?

B: Three yuan each.

A: That's a bargain. Here you are.

B: All right. Welcome again next time.

A: 你卖的小吃看起来好诱人,我要一个。

B: 稍等,这就给你现炸一个。

A: 你这个小吃叫什么名字?

B: 耳朵眼炸糕。

A: 真有趣,这是怎么制作的呢?

B: 它的工序很复杂。首先要将精选糯米经水泡胀后磨成浆状,加入水和碱发酵后作为面皮,包入红豆馅,温油下锅,勤翻动,炸熟后出锅。

A: 听起来就很好吃。

B: 炸好了,这是你的耳朵眼炸糕。

A: 多少钱?

B: 三元一个。

A: 真便宜,给你钱。

B: 好的,欢迎下次再来。

生词注解 Notes

① dynasty /ˈdɪnəstɪ/ *n.* 王朝；朝代

② reflect /rɪˈflekt/ *vt.* 反映；反射

③ nourishing /ˈnʌrɪʃɪŋ/ *adj.* 有营养的

④ invigorating /ɪnˈvɪɡəreɪtɪŋ/ *adj.* 精力充沛的；生机勃勃的

⑤ affordable /əˈfɔːdəbl/ *adj.* 负担得起的

⑥ adhere /ədˈhɪə/ *vi.* 坚持

⑦ tempting /ˈtemptɪŋ/ *adj.* 吸引人的；诱惑人的

⑧ fermentation /ˌfɜːmenˈteɪʃn/ *n.* 发酵

⑨ alkali /ˈælkəlaɪ/ *n.* 碱；可溶性无机盐

饮食文化

桂发祥十八街麻花

Guifaxiang 18th Street Fried Dough Twists

 导入语　Lead-in

　　桂发祥十八街麻花因店铺坐落于天津大沽南街十八街而得名。"桂发祥"的寓意是"桂子飘香、发愤图强、吉祥如意"。在浩如烟海的历史长卷上，任何一种物体都是见证一个时代的符号，小小的麻花也承载了时代发展变迁的历史印迹。在全国首届名小吃认定会上，桂发祥十八街麻花被认定为"中华名小吃"，其传统的制作技艺被评为天津市非物质文化遗产，并入选《国家级非物质文化遗产名录》。每个桂发祥麻花中间都夹有一根由芝麻、桃仁、瓜子仁、青梅、桂花、青红丝等小料配制而成的什锦馅酥条，馅溢不散、酥脆香甜、久放不绵。

文化剪影 Cultural Outline

Guifaxiang 18th Street Fried Dough Twists is regarded as top among the traditional Tianjin **snacks**①, which has become the symbol of Tianjin cuisine and a wonderful work of Chinese cuisine.

桂发祥十八街麻花被誉为"津门首绝",已经成为天津美食的象征,是中华美食中的一朵奇葩。

Guifaxiang 18th Street Fried Dough Twists **emerged**② at the end of the 19th century, hand-made and **preservative**③-free. It can be stored for three months in spring and autumn, and two months in summer.

桂发祥十八街麻花出现于19世纪末,纯手工制作,不含防腐剂。在春秋两季可以储存三个月,夏季可以储存两个月。

Guifaxiang 18th Street Fried Dough Twists is characterized by mixing with ten kinds of **stuffing**④. The choice of materials is **exquisite**⑤, the color is **appetizing**⑥ and the shape is even and full.

桂发祥十八街麻花的特色是将什锦馅料融入麻花中,其选料考究、颜色诱人、均匀饱满。

饮食文化

佳句点睛 Punchlines

1. Guifaxiang 18th Street Fried Dough Twists combines traditional production with modern food technology.

桂发祥十八街麻花将传统生产与现代食品工艺相结合。

2. Guifaxiang 18th Street Fried Dough Twists integrates traditional culture with modern fashion elements.

桂发祥十八街麻花将传统文化与现代时尚元素进行融合。

3. Guifaxiang 18th Street Fried Dough Twists focuses on the combination of **nutrition**[7] and health.

桂发祥十八街麻花注重营养与健康的结合。

情景对话 Situational Dialogue

A: Tianjin is a beautiful city, right?

B: Yes, there're also lots of delicacies, like Go Believe or something.

A: I'm hungry. Let's get something to eat.

B: Me too. You see, there're many people in front of that shop.

A: Let's line up there, for the line is so long. I guess the food is delicious.

B: Look! It is Guifaxiang 18th Street Fried Dough Twists, which sounds interesting.

A: It is one of the three **specialties**⑧ of Tianjin. I'll surely have a try in person.

B: I'm afraid there's too much sugar in it. I can't have too much sweet food.

A: Don't worry. We can buy some without sugar.

B: Oh, yes. They are yummy.

A: I am very interested in the making process. Maybe we can visit the factory.

B: OK, let's go.

A: 天津真是一座美丽的城市,对吧?

B: 对,还有好多美味佳肴,比如狗不理包子什么的。

A: 说得我都饿了,我们去吃点东西吧。

B: 我也是。你看,那家商店前面有很多人。

A: 咱们在那里排队吧,队伍好长,我猜卖的东西肯定很好吃。

B: 是桂发祥十八街麻花,听起来很有趣。

A: 这是天津的三大小吃之一,我一定要亲口尝尝。

B: 我担心它太甜了。我不能吃太多的甜食。

A: 别担心。我们可以买一些不加糖的。

B: 噢,是的。它们真好吃。

A: 我对制作过程很感兴趣,也许我们可以参观一下工厂。

B: 行,我们走吧。

饮食文化

生词注解 Notes

① snack /snæk/ n. 小吃;点心

② emerge /ɪˈmɜːdʒ/ vi. 浮现;暴露

③ preservative /prɪˈzɜːvətɪv/ n. 防腐剂

④ stuffing /ˈstʌfɪŋ/ n. 配料

⑤ exquisite /ɪkˈskwɪzɪt/ adj. 精致的;细腻的

⑥ appetizing /ˈæpɪtaɪzɪŋ/ adj. 开胃的;促进食欲的

⑦ nutrition /njuˈtrɪʃn/ n. 营养;营养品

⑧ specialty /ˈspeʃəltɪ/ n. 特产;专长

德州扒鸡

Dezhou Braised Chicken

导入语 Lead-in

德州扒鸡是山东省"德州三宝"之一,由烧鸡演变而来,制作人偶然烹调得到诀窍,又经后人不断改进技法,以中药材入食,采用多种烹调手段制作而成。随着德州交通地位的日益显著,德州扒鸡的名声也日益响亮。早在清朝乾隆年间,德州扒鸡就被列为山东贡品,为皇室贵族享用。在当代,德州各家制作扒鸡的技艺相互融合,并在美食评鉴大赛上拔得头筹。德州扒鸡不仅是山东名吃,而且承载着德州交通发展的历史文化印记,被列入《国家非物质文化遗产名录》,堪称"中华第一鸡"。

文化剪影 Cultural Outline

Dezhou **Braised**① Chicken, founded in the Qing Dynasty, was cooked by a merchant named Jia by chance and won praises, later he devoted himself to developing it. As Dezhou was located at the heart of the railroad system, it spread all around of China along the railway in the Republic of China. Then Dezhou Braised Chicken constantly improved skills and has become famous in China nowadays.

德州扒鸡始创于清朝，由贾姓商人偶然烹调，获得称赞，随后他又潜心研究，不断发展改进；传至民国，由于德州当时地处铁路交通枢纽，这道美食随铁路传向大江南北；德州扒鸡不断提高制作工艺，享誉中国。

Dezhou Braised Chicken can be called the representative of Shandong cuisine, the chicken skin is bright and ruddy, and the meat is tender and full of fragrance and nourishing. It is a classic representative of Chinese cuisine mixed with traditional Chinese medicine, as well as a popular folk delicacy.

德州扒鸡堪称鲁菜代表，色泽红润，肉质软嫩，五香入味，营养滋补，既是中药材入食的经典代表，又是广受欢迎的民间美食。

Dezhou Braised Chicken has unique features in color, aroma, taste and shape. It mixes the methods of frying, smoking, **marinating**② and braising, and adds a variety of traditional Chinese medicine that can **tonify**[3] stomach and kidney and be easy to digestion.

德州扒鸡色、香、味、型均有独到之处,因制作时融合了炸、熏、卤、烧的方法,添加了多味健脾开胃的中药,故具有健胃、补肾、助消化等功能。

佳句点睛 Punchlines

1. Dezhou Braised Chicken has always been a regular delicacy of Dezhou people.

德州扒鸡一直是德州人餐桌上的常客。

2. Dezhou Braised Chicken is cooked together with the traditional Chinese medicine by a variety of cooking methods.

德州扒鸡以中药材入食,采用多种烹调手段制成。

3. Dezhou Braised Chicken has been a traditional flavor of Dezhou. When eating it hot, diners simply hold and shake it, and the meat separates from the bones.

作为德州的传统风味,德州扒鸡热吃时,食客只需随手撕扯,骨肉便即刻分离。

情景对话 Situational Dialogue

A: Good afternoon. May I help you?

B: Well, thank you. This is my first time to Dezhou. Do you have any recommendation?

A: Dezhou Braised Chicken is a specialty, which is famous for its pure taste and tender meat. Here is the picture.

B: It looks very nice. Can you tell me how it is made?

A: We selected about one kilogram of local **cockerels**④ or small hens without laying eggs, cleaned their internal organs, dipped the chicken in cold water to fix the shape and mixed with sugar color, later deep-fried it for two minutes, added the spices and seasoning, and **simmered**⑤ it for three to four hours.

B: It's so **complicated**⑥. It must taste good. I'll have one, please.

A: OK. Anything else?

B: A sweet and sour **carp**⑦ and **shredded**⑧ potato in vinegar.

A: Would you like something to drink?

B: A glass of Qingdao Beer.

A: OK, just a moment.

B: All right.

A: 下午好,我能帮你吗?

B: 是的,谢谢。我是第一次到德州,你有什么推荐吗?

A: 我们这里的德州扒鸡是远近闻名的特色菜,味道纯美、肉质软嫩。你看,这是图片。

B: 看起来很不错,你能告诉我这道菜是怎么制作的吗?

A: 我们选用一公斤左右的当地小公鸡或未下蛋的小母鸡,掏净内脏、冲洗干净。将鸡浸入冷水中做好造型,涂抹糖色,放入锅中先用油炸两分钟,再用香料调料,入水焖煮三到四小时后即可出锅。

B: 这么复杂啊,味道一定不错。请给我来一份。

A: 好的,还需要什么吗?

B: 再要一个糖醋鲤鱼和醋熘土豆丝。

A: 饮品有什么需要的吗?

B: 来一杯青岛啤酒。

A: 好的,请稍等。

B: 好。

生词注解 Notes

① braise /breɪz/ vt. 炖;焖

② marinate /ˈmærɪneɪt/ vt. 腌制;浸泡

③ tonify /ˈtəʊnɪfaɪ/ vt. 滋补;补益

④ cockerel /ˈkɒkərəl/ n. (未满一年的)小公鸡

⑤ simmer /ˈsɪmə(r)/ vi. 煨;(水或食物)慢慢沸腾

⑥ complicated /ˈkɒmplɪkeɪtɪd/ adj. 复杂的;难懂的

⑦ carp /kɑːp/ n. 鲤鱼;鲤科属鱼

⑧ shredded /ˈʃredɪd/ adj. 切成丝的;切碎的

饮食文化

肉夹馍

Marinated Meat in Baked Bun

导入语　Lead-in

肉夹馍是陕西省著名的特色小吃之一，得名于古汉语"肉夹于馍"。陕西地区有使用白吉馍的腊汁肉夹馍、宝鸡西府的肉臊子夹馍（肉臊子中放食醋）、潼关的潼关肉夹馍（与白吉馍不同，其馍体焦黄，纹理清晰，皮酥里嫩，老潼关肉夹馍是热馍夹凉肉，饼酥肉香，爽而不腻）。肉夹馍合肉、馍为一体，互为烘托，将各自滋味发挥到极致，馍香肉酥，令人回味无穷。吃肉夹馍的正宗姿势为水平持馍，从两侧咬起。水平持馍可以使腊汁肉的肉汁充分浸入馍中，不致溢出。肉夹馍颇似西方的汉堡包，所以又有"Chinese hamburger"之称。

第三部分 特色美食

文化剪影 Cultural Outline

According to the historical record, the state of Han located in the **triangle**① zone of Shaanxi, Shanxi and Henan could make **marinated**② meat in the Warring States Period. After the state of Han was wiped out by the state of Qin, the way to make marinated meat was spread to Qin. The marinated soup used to cook Chinese hamburger has been handed down from generation to generation.

据史料记载,战国时期,位于秦晋豫三角地带的韩国已能制作腊汁肉,秦灭韩后,制作工艺传入秦国。制作腊汁肉的陈汤也一代代流传下来。

Marinated Meat in Baked Bun with marinated meat is tasty but not greasy, tender and delicious. Baiji Bun is crisp outside and tender inside, which is the best partner of the marinated meat.

腊汁肉夹馍,其腊汁香而不腻,肉质肥美柔软。白吉馍皮薄松脆,内里绵软,是腊汁肉的最佳拍档。

The making process of Marinated Meat in Baked Bun is as follows that we can select the **superior**③ meat to boil with more than twenty ingredients such as salt, ginger, green onion, clove, **cinnamon**④, rock sugar and so on.

腊汁肉的做法是选用上等硬肋肉,用盐、姜、葱、丁香、桂皮、冰糖

113

等二十多种调料汤一起熬煮。

佳句点睛 Punchlines

1. Marinated Meat in Baked Bun is crispy and yummy, making diners linger on and enjoy the endless aftertaste.

肉夹馍酥脆可口,令食者流连忘返,回味无穷。

2. In its long-term development, Marinated Meat in Baked Bun has formed different flavors and characteristics in different regions.

在长期的发展中,肉夹馍在不同地域形成了不同的风味与特色。

3. Marinated Meat in Baked Bun is not only yummy, but also loved by many **domestic**[5] and foreign tourists.

肉夹馍不仅美味可口,而且深受海内外众多游客喜爱。

情景对话 Situational Dialogue

A: Where did you go on the Mid-autumn Festival?

B: I went to Shaanxi for a trip.

A: Wow, it must be very interesting.

B: Yeah, I tasted many **delicacies**[6] there.

A: Do you have anything that impressed you most about the food?

B: Of course, Marinated Meat in Baked Bun is my favorite.

A: Why do you like it?

B: Because it is yummy and the meat is tasty but not **greasy**⑦, filling you with **lingering**⑧ fresh flavor.

A: Do you know its origin?

B: According to the legend, it was made in the Warring States Period with a history of thousands of years.

A: How was it made?

B: Baiji Buns with marinated meat, which is boiled with more than twenty **ingredients**⑨.

A: That does sound good, so I must taste it when I get a chance.

A: 你中秋节去哪里了?

B: 我到陕西旅游去了。

A: 哇,那一定很有趣吧。

B: 是的,我品尝了很多当地美食。

A: 有什么让你记忆深刻的食物吗?

B: 当然了,我最喜欢吃的是肉夹馍。

A: 你为什么喜欢它?

B: 因为它非常好吃,肉质肥而不腻,让人满口生香,余味悠长。

A: 你知道它的起源吗?

B: 相传战国时期肉夹馍就已出现,距今已有几千年的历史。

A: 它是如何制作的呢?

B: 白吉馍配腊汁肉,腊汁肉是由二十多种调料汤煮成的。

A: 听起来确实不错,有机会我一定要品尝一下。

生词注解 Notes

① triangle /ˈtraɪæŋgl/ *n.* 三角形；三角形物体

② marinated /ˈmærɪneɪtɪd/ *adj.* 使……受腌汁浸泡的；腌泡的

③ superior /suːˈpɪərɪə(r)/ *adj.* 优秀的；出色的

④ cinnamon /ˈsɪnəmən/ *n.* 肉桂；肉桂色

⑤ domestic /dəˈmestɪk/ *adj.* 国内的；家庭的

⑥ delicacy /ˈdelɪkəsi/ *n.* 美味佳肴；山珍海味

⑦ greasy /ˈgriːsi/ *adj.* 油腻的；含脂肪多的

⑧ lingering /ˈlɪŋgərɪŋ/ *adj.* 绵长的；挥之不去的

⑨ ingredient /ɪnˈgriːdɪənt/ *n.* 原料；要素

凉皮

Steamed Cold Rice Noodles

 导入语 Lead-in

凉皮是陕西的一种地方特色小吃,相传源于秦汉时期,距今已有两千多年的历史。传说有一年陕西户县秦镇一带大旱,稻谷枯萎,百姓无法向朝廷纳供大米,一位名叫李十二的人用大米碾成面粉,蒸出面皮,献给秦始皇,秦始皇吃后大悦,命人每天制作供他食用,由此便形成了久负盛名的汉族传统小吃——秦镇大米面皮子,也就是凉皮。此后,凉皮成为当地每年进贡皇帝的贡品。凉皮具有筋、薄、细、穰、辣、咸、香和爽的特色,因此受到大众的普遍欢迎,食用时让人口感凉爽,回味无穷。

文化剪影　Cultural Outline

Steamed Cold Rice Noodles, one of the special dishes of Shaanxi Province, is the general designation of Ganmianpi, Mianpi, Mipi and Niangpi. It is always offered together with wheat **gluten**①, vinegar, salt, chilli oil, **mashed**② garlic in cool water and bean **sprout**③.

凉皮是陕西特色小吃之一,是擀面皮、面皮、米皮和酿皮的统称,常与面筋、醋、盐、辣椒油、蒜汁和豆芽一起凉拌食用。

Steamed Cold Rice Noodles has a long history, now spreading to many other places in China. Liangpi is mostly made of wheat flour and rice flour, as the method of making is different, the flavor is also different from each other. There're mainly Majiang Liangpi, Qinzhen Mipi and Hanzhong Mianpi, and so on.

凉皮具有悠久的历史,现已传遍中国的大江南北。凉皮多由小麦面粉和米粉制作而成,因制作方法各异,口味也各具特色。主要有麻酱凉皮、秦镇米皮和汉中面皮等。

Shaanxi Steamed Cold Rice Noodles has long enjoyed a fame as a special snack, not only because it has a long history, but also because of its delicacy, **affordability**④ and **vitality**⑤.

作为特色小吃,陕西凉皮久负盛名,不仅因为它具有悠久的历史,而且好吃实惠,具有旺盛的生命力。

佳句点睛 Punchlines

1. The good taste is the main reason why Steamed Cold Rice Noodles is loved by so many people.

口感好是凉皮受到许多人喜爱的主要原因。

2. Eating fast is another important reason why Steamed Cold Rice Noodles is popular.

就餐快是凉皮受欢迎的另一个重要原因。

3. Steamed Cold Rice Noodles is a natural green **pollution-free**⑥ food suitable for all seasons.

凉皮可谓是四季皆宜的天然绿色无公害食品。

情景对话 Situational Dialogue

A: What is your favorite Chinese dish?

B: Like most foreigners, I really like Sweet & Sour Spareribs and Kung Pao Chicken (Spicy Diced Chicken with Peanuts).

A: Have you ever tried Shaanxi Steamed Cold Rice Noodles?

B: No, I have not, but I have heard about it.

A: Yeah, it is famous throughout the country, with a long history. If you have time, I can introduce you more about Liangpi.

B: Oh, that would be great.

A: The major ingredients of it are rice and minor **ingredient**⑦ made from seasonings. It has a sense of tender.

B: Rice?

A: Oh, we rinse the rice in advance, and **grind**⑧ into rice milk after **soaking**⑨, and then steam the rice milk by special tools.

B: How can you think of such a complicated way to make it?

A: We eat the same food in different ways.

B: Chinese people are really smart!

A：你最喜欢的中国菜是什么？

B：像大多数外国人一样，我非常喜欢糖醋排骨和宫保鸡丁。

A：你吃过陕西凉皮吗？

B：没有，但我听说过。

A：凉皮历史悠久，全国有名。如果你有时间，我可以给你介绍一下凉皮。

B：噢，那太好了。

A：凉皮主要是由大米和其他时令配菜制成，吃起来口感软糯。

B：大米？

A：噢，我们事先把米洗干净，浸泡后磨成米浆，然后用专用工具蒸米浆。

B：你们怎么能想到这么复杂的制作方法呢？

A：我们对同样的食物有不同的吃法。

B：中国人真的很聪明！

生词注解 Notes

① gluten /ˈgluːtn/ *n.* 面筋;蛋白谷

② mashed /mæʃt/ *adj.* 捣碎的;被捣成糊状的

③ sprout /spraʊt/ *n.* 芽;萌芽

④ affordable /əˈfɔːdəbl/ *adj.* 价格实惠的

⑤ vitality /vaɪˈtælətɪ/ *n.* 活力;生气

⑥ pollution-free /pəˈluːʃənfriː/ *adj.* 无污染的

⑦ ingredient /ɪnˈgriːdɪənt/ *n.* 原料;要素

⑧ grind /graɪnd/ *vt.* 磨碎;研磨

⑨ soak /ˈsəʊk/ *vt.* 浸泡;浸入

饮食文化

羊肉泡馍

Pita Bread Soaked in Lamb Soup

导入语　Lead-in

羊肉泡馍简称"羊肉泡",是以卤羊肉、烤馍、粉丝、黄花菜、木耳、糖蒜、香菜等为原料制作而成的传统陕西名吃。羊肉泡馍古称"羊羹",源于唐朝士兵为改善干馍口味,将羊肉汤同干馍一起烹煮的做法。羊肉泡馍制作精细,讲究汤浓、肉烂、馍香,营养丰富,香飘十里,食后人微微出汗、通体舒畅、回味无穷。也因其暖胃、耐饥、补肾,故适合在气候寒冷时食用。随着文化的传播,羊肉泡馍吸引了五湖四海的食客,已经成为西安美食的代名词。

文化剪影 Cultural Outline

Pita Bread Soaked in Lamb Soup is said to have evolved from the ancient "Yanggeng" (Lamb Soup). In the Tang Dynasty, Tuotuo Buns were used by soldiers for marching. To improve the taste of dry **rations**[①] and make it easier swallow, soldiers cooked it with lamb soup, which was the **prototype**[②] of Pita Bread Soaked in Lamb Soup.

相传羊肉泡馍是在古代"羊羹"的基础上演化而来的。盛唐时,饦饦馍被唐朝士兵用于行军干粮。为改良干粮口味、方便下咽,士兵将其与羊肉汤一同烹煮,这就是羊肉泡馍的雏形。

As the representative of Shaanxi cuisine, the Pita Bread Soaked in Lamb Soup has been praised all the time by its lamb and thick soup, fresh and **mellow**[③] taste, fragrant smell and endless aftertaste.

羊肉泡馍作为陕西美食的代表,因肉烂汤浓、咸鲜入味、口味醇厚、香气扑鼻、回味无穷,一直为人称道。

The complicated and **delicate**[④] making process, the delicious and mellow taste experience, the lingering sensory enjoyment and the **therapeutic**[⑤] effect of warming the stomach and nourishing the body make the Pita Bread Soaked in Lamb Soup different from other rough-featured northwest food made of flour.

复杂精细的制作流程、鲜美醇厚的味觉体验、回味无穷的感官享

受和暖胃滋补的食疗功效,使羊肉泡馍有别于其他粗犷的西北面食。

佳句点睛 Punchlines

1. Pita Bread Soaked in Lamb Soup has become the national delicacy with its mellow and delicious taste.

羊肉泡馍因其醇厚鲜美的味道成为闻名全国的美食。

2. The method of mixing staple food and dishes is also a unique representative of northern Shaanxi food made of flour.

将主食与菜肴混合烹煮的烹调法是陕北面食的特色。

3. Pita Bread Soaked in Lamb Soup requires the cooperation between **chef**⑥ and guest and strict eating method.

羊肉泡馍要求厨客协作,食用方法严格。

情景对话 Situational Dialogue

A: May I help you, sir?

B: Yes, I'd like to try your special dish — Pita Bread Soaked in Lamb Soup.

A: OK, just a moment, please.

A: Sir, please break the bread into small pieces about the size of **soybean**⑦ and put them in the bowl while waiting for your meal. Then

I will send the bread to cook for you later.

B: What a unique process. Can you tell me how this dish is made?

A: Of course, firstly boil the high-quality lamb with onion, ginger, Sichuan pepper and other seasonings over low heat for twelve hours, then strain out the stock for later use. Mix the bread, broth, lamb with cabbage and other accessories, add salt, cooking wine and other seasonings before it is done. Eventually, we'd better add **coriander**® to have it.

B: It's so complicated. It must be yummy. I can't wait to enjoy it.

A: Take you time, the bread should be in small pieces, so that our chef can raise its freshness and bring you the most delicious pita bread soaked in mutton soup. Would you like anything else?

B: No, thank you.

A：先生,有什么需要吗?

B：是的,我想尝尝你们这里的招牌美食——羊肉泡馍。

A：好的,请您稍等。

A：先生,请您在等餐期间将这个烤馍掰成黄豆大小的小块,放在碗中,等会儿我会将您掰好的馍送给厨师烹煮。

B：好独特的制作流程!你能告诉我这道菜是怎么制作的吗?

A：当然可以,先将优质羊肉同葱、姜、花椒等各种佐料用小火熬煮十二小时,滤出高汤备用。将掰好的馍与高汤、原肉混合熬煮,配以白菜、粉丝等辅料,出锅前加入盐、料酒等调味,最后配以香菜提鲜,即可享用。

B: 这么复杂啊。一定很好吃。我都迫不及待了。

A: 先生,不用急,烤馍一定要掰得小一点,这样我们的厨师才能呈上最美味的羊肉泡馍。请问还需要别的菜吗?

B: 不需要了,谢谢你。

生词注解 Notes

① rations /ˈræʃnz/ *n.* 口粮;定量食物

② prototype /ˈprəʊtətaɪp/ *n.* 原型;标准

③ mellow /ˈmeləʊ/ *adj.* 圆润的;柔和的

④ delicate /ˈdelɪkət/ *adj.* 精美的;雅致的

⑤ therapeutic /ˌθerəˈpjuːtɪk/ *adj.* 治疗的;有益健康的

⑥ chef /ʃef/ *n.* 厨师;大师傅

⑦ soybean /ˈsɔɪbiːn/ *n.* 大豆;黄豆

⑧ coriander /ˌkɒriˈændə(r)/ *n.* 芫荽;香菜

臊子面

Saozi Minced Noodles

 导入语 Lead-in

臊子面是陕西的风味小吃，以宝鸡的岐山臊子面最为正宗，在陕西关中平原和甘肃陇东等地广为流行，其特点是面条细长、厚薄均匀、臊子鲜香、面汤油光红润、味道浑厚而不腻。臊子面品种繁多，有手工面和机制面，有荤也有素，但最有名的还是岐山臊子面。所谓臊子面，就是用肉末做浇头的面条，臊子是将生猪肉切片，加酱油、精盐、红醋、辣椒粉等制成，然后将面粉、温水、碱和成面团再揉成长方块，擀薄后切成细条，入锅内煮两滚后捞入冷水盆中，再放上肉臊子，最后浇酸汤。臊子面的特点是"酸辣香，薄筋光，煎稀汪"。所谓"煎""汪"，就是面条热烫、油多。

文化剪影 Cultural Outline

Saozi Minced Noodles is a famous traditional noodle in Shaanxi. It is said that it evolved from "**Longevity**① Noodles" in the Tang Dynasty, which has become a good dish to serve the guests in celebrating the birthday of the elders and children and other festivals.

臊子面是陕西著名的传统面食。据说是由唐朝的"长寿面"演化而来,是老人寿辰、小孩生日及其他节日的待客佳品。

Saozi Minced Noodles is unique in Shaanxi Province. Its uniqueness is mainly reflected in the way it is made and eaten. In the process of making it, the noodles and Saozi needs to be cooked separately. When enjoying Saozi Minced Noodles, we should **adhere to**② the custom of "only eating noodles instead of drinking soups".

臊子面在陕西面食中独领风骚,它的独特性主要体现在面的做法和吃法上。在制作臊子面的过程中,面条和臊子需要分开做、分开煮;在食用臊子面的时候,应该秉承"只吃面不喝汤"的风俗习惯。

Saozi Minced Noodles can be popular among the people, evolving into a unique classic food made of flour benefiting from the neighborhood **mutual**③ assistance and the harmonious **coexistence**④ of fine tradition.

臊子面能在民间久盛不衰,并演变为独具特色的经典面食,得益于邻里之间互帮互助、和谐相处的优良传统。

佳句点睛 Punchlines

1. Noodles are one of the essential **staple**⑤ foods in Chinese cuisine.
面条是中国美食里重要的主食之一。

2. Qishan Saozi Minced Noodles, also known as Qishan Noodles, is made of superior materials.
岐山臊子面又称为"岐山面",选料讲究,制作精良。

3. Local people also serve Saozi Minced Noodles as their delicacy during festivals and birthdays.
当地人民在过节、做寿时也把臊子面作为待客的美食。

情景对话 Situational Dialogue

A: My dear, what's for supper? I'm starving.

B: Oh, I didn't prepare. Do you have any good ideas?

A: We'd better have some vegetables and grains, which will benefit our health.

B: I think so. But it's also very important to eat some meat for the sake of nutrition balance.

A: Yes, exactly! Is there any food that will balance the two points?

B: Let me see. I know Saozi Minced Noodles can balance such two points.

A: Do you know how to make the food?

B: Just prepare meat, eggs and vegetables for me.

A: OK. I can help you. Here are the ingredients for you.

B: It's done, just have a try.

A: Well, it's so delicious.

B: I'm glad you enjoy it.

A: 亲爱的,晚饭吃什么? 我快饿死了。

B: 噢,我没有准备。你有什么好主意吗?

A: 我们最好吃点儿蔬菜和谷物,这对健康很有好处。

B: 我也这么认为。但是,为了营养平衡,吃一些肉也很重要。

A: 是的,完全正确! 有什么食物能平衡这两点的吗?

B: 让我看看。我知道臊子面可以平衡这两点。

A: 你知道怎么做臊子面吗?

B: 请给我准备些肉、蛋和蔬菜。

A: 好的,我可以帮你。给你需要的配料。

B: 做好了。请试试吧。

A: 嗯,太好吃了。

B: 你喜欢就好。

生词注解 Notes

① longevity /lɒnˈdʒevətɪ/ *n.* 长寿

② adhere /ədˈhɪə(r)/ to 坚持

③ mutual /ˈmjuːtʃuəl/ *adj.* 共同的；相互的

④ coexistence /ˌkəʊɪɡˈzɪstəns/ *n.* 共存；和平相处

⑤ staple /ˈsteɪpl/ *n.* 主食；主要产品

胡辣汤

Hot Spicy Soup

 导入语 Lead-in

胡辣汤是地道的河南小吃，它醇厚的秘密藏在由几十种香辛料调制而成的骨汤里。正宗的胡辣汤中只用胡椒，不用辣椒。辛勤劳作的中原儿女在几百年的摸索中总结出了这些香辛料的配比，在保证口感的基础上又各有特色。除了牛羊肉以外，面筋、黑木耳、黄花菜、粉条和花生等配菜也常常出现在胡辣汤里。一碗热腾腾的胡辣汤端上桌，既有色泽浓郁的汤底又有营养丰富的食材。胡辣汤已经成为中原的一张名片，它正以独特的香味和醇厚的口感吸引着越来越多的食客。2021年，胡辣汤被国务院列入《第五批国家级非物质文化遗产代表性项目名录》。

文化剪影 Cultural Outline

The preparation of traditional Hot Spicy Soup starts at three or four o'clock in the morning, from preparing ingredients to several hours of cooking, and finally the rich and appetizing Hot Spicy Soup is done.

传统胡辣汤的制作要从凌晨三四点钟开始,从准备食材到长达数小时的烹调,最终制作出味香可口的胡辣汤。

The perfect **integration**① of food ingredients and soup base is embodied in the unique central plains cuisine — Hot Spicy Soup, which is also characterized by the **fusion**② of all directions in the Central Plains Region.

食材和汤底的完美融合,也让中原地区包容万象、融汇四方的特点体现在胡辣汤这道独特的中原美食上。

Hot Spicy Soup is reasonable collocation and rich in nutrition. And from the perspective of traditional Chinese medicine, Hot Spicy Soup has the effect of **appetizing**③, and **dispelling**④ the wind and chills.

胡辣汤搭配合理,营养丰富。从中医角度来说,胡辣汤还有消食开胃、祛风散寒的功效。

饮食文化

佳句点睛 Punchlines

1. Hot Spicy Soup is full of rich taste and nutrition.

胡辣汤味道浓郁，营养丰富。

2. Hot Spicy Soup has a long history and profound cultural heritage.

胡辣汤历史悠久，文化底蕴深厚。

3. The authentic Hot Spicy Soup is characterized by the sour and spicy taste.

正宗胡辣汤的口味特征是酸和辣。

情景对话 Situational Dialogue

A: Welcome back home, Mr. Wei!

B: I've been abroad for quite a long time, but what I miss most is such a Hot Spicy Soup.

A: It seems that you like soup very much.

B: The habit of Henan people like drinking soup probably has something to do with the **migration**⑤ of **refugees**⑥ in history. Soup can **satisfy**⑦ hunger best in the process of moving from one place to another.

A: Why do you prefer Hot Spicy Soup?

B: Because of its special flavor, the making process of Hot Spicy Soup is not a simple thing, and it's needed to use nearly thirty kinds of Chinese medicinal materials such as Chinese prickly ash, black pepper, and so on.

A: There're so many spices in the soup.

B: Yes, but more important is the taste of Hot Spicy Soup warms the heart of many people who are away from home.

A: The making process of Hot Spicy Soup is required to put enough ingredient, which is similar to be the real man in the society.

B: A bowl of Hot Spicy Soup looks simple, but it has reflected the feelings and character of Henan people.

A: Mr. Wei, let me take you to have a bowl of Hot Spicy Soup!

B: OK!

A：魏老师，欢迎回家！

B：在国外这么长时间，最想念的就是这口胡辣汤。

A：看来魏老师很喜欢喝汤啊！

B：河南人爱喝汤的习惯大概和历史上的流民迁徙有关，颠沛流离中汤最能充饥。

A：您为什么偏爱胡辣汤？

B：因为它风味独特，而且熬制一碗胡辣汤也不是简单的事儿，要用花椒、胡椒等近三十种中药材，才能成就一碗正宗的胡辣汤。

A：这么多调料啊！

B：是啊，更重要的是胡辣汤的味道温暖了游子们的心。

饮食文化

A: 胡辣汤和做人一样，都得实实在在。

B: 一碗胡辣汤看似简单，却体现了河南人的情怀和性格。

A: 魏老师，我带您先去喝汤吧。

B: 中！

 生词注解 Notes

① integration /ˌɪntɪˈgreɪʃn/ n. 集成；综合

② fusion /ˈfjuːʒn/ n. 融合；合并

③ appetizing /ˈæpɪtaɪzɪŋ/ adj. 开胃的；促进食欲的

④ dispel /dɪˈspel/ vt. 驱散；消除（烦恼等）

⑤ migration /maɪˈgreɪʃn/ n. 迁移；移民

⑥ refugee /ˌrefjuˈdʒiː/ n. 难民；避难者

⑦ satisfy /ˈsætɪsfaɪ/ vt. 满足；使……相信

刀削面

Sliced Noodles

 导入语 Lead-in

山西地处黄河中游,是世界农业起源中心之一,也是中国面食文化的发祥地,其面食文化历史悠久、内涵丰富、积淀深厚,令人瞩目。山西面食种类繁多,其中以大同的刀削面最有名,堪称一绝。刀削面全凭刀削,因此得名,以刀功和削技的绝妙而被称为"飞刀削面"。用刀削出的面叶中厚边薄、棱锋分明、形似柳叶,入口外滑内筋、软而不粘,深受欢迎。刀削面物美价廉,色、香、味俱全,因其独特的风味和特殊的制作过程,影响也越来越广泛。

饮食文化

文化剪影 Cultural Outline

China has gathered the **essence**① of the world's pasta while Shanxi is the **prosperous**② place of Chinese pasta. As one of the representatives of traditional pasta of the Han nationality, Shanxi pasta has a wide variety and a unique taste while Sliced Noodles can be called "The King of Noodles."

世界面食在中国，中国面食在山西。作为汉族传统面食的代表之一，山西面食种类繁多、味道独特，刀削面堪称"面食之王"。

Sliced Noodles bears the human **sentiment**③ and local feeling. People's emotion to Sliced Noodles is beyond a bowl of pasta, but their **attachment**④ and aftertaste to their hometown.

刀削面寄托着人情和乡情，人们对刀削面的情感已经超越了一碗面食，而是寄托了对家乡的依恋和回味。

Sliced Noodles **embodies**⑤ folk wisdom and can be called exquisite works of art, from kneading dough, slicing noodles to making Saozi are all of strict **requirements**⑥.

刀削面体现了民间智慧，可谓是一种精美的艺术品。从和面、削面到做臊子都有严格的要求。

佳句点睛 Punchlines

1. The pasta culture is a cultural **phenomenon**[7] arising from the development of noodles.

面食文化是由面食发展而来的一种文化现象。

2. Saozi is the key to make Sliced Noodles, which is featured by various types and bright color.

臊子是做刀削面的关键,其种类多样,色彩鲜艳。

3. Sliced Noodles is **convenient**[8] and fast delicacy.

刀削面是一种方便快捷的佳肴。

情景对话 Situational Dialogue

A: Hi, Jane.

B: Hello, Lily. Today is my birthday. I wanna invite you to have a dinner. Let's dine out together.

A: That's great. It's my pleasure.

B: Let's have a look at the menu. I can't read these Chinese dishes. What do you recommend to me?

A: There're a lot of Chinese delicacies. But we'd better eat some noodles as today is your birthday.

B: Why should we have noodles?

A: Because the noodle has special meaning, which can bless you a long life.

B: It's funny. Thank you for sharing with me. What kind of noodles should we have?

A: What about Sliced Noodles? It's totally different from other noodles.

B: OK. Chinese food is different from western food, but they have something in common.

A: Yes, for example, pizza is similar to Chinese pie while **spaghetti**⑨ is similar to noodles.

B: That's right.

A: 你好,简。

B: 你好,莉莉。今天是我的生日,我想请你吃饭。我们一起出去吃饭吧。

A: 太好了,很荣幸。

B: 来看看菜单吧。我看不懂这些中国菜,你有什么推荐的吗?

A: 中国美食有很多。但是我们最好吃一些面条,因为今天是你的生日。

B: 我们为什么要吃面条?

A: 因为面条有特殊意义,可以保佑你长寿。

B: 真有趣,谢谢分享。我们应该吃什么样的面条?

A: 刀削面怎么样?它和其他面条完全不同。

B: 好的。中国菜不同于西方菜,但它们也有相通之处。

A: 是的,比如说披萨和中国馅饼相似,而意大利面则和中国面条相似。

B: 说得对。

生词注解 Notes

① essence /ˈesns/ n. 本质;实质

② prosperous /ˈprɒspərəs/ adj. 繁荣的;兴旺的

③ sentiment /ˈsentɪmənt/ n. 感情;情绪

④ attachment /əˈtætʃmənt/ n. 依恋;附件

⑤ embody /ɪmˈbɒdɪ/ vt. 体现;使……具体化

⑥ requirement /rɪˈkwaɪəmənt/ n. 要求;必要条件

⑦ phenomenon /fəˈnɒmɪnən/ n. 现象;奇迹

⑧ convenient /kənˈviːnɪənt/ adj. 方便的;适当的

⑨ spaghetti /spəˈɡetɪ/ n. 意大利面

太原头脑

Soup with Eight Ingredients

导入语 Lead-in

深秋以后，天气渐凉，山西太原人早上习惯喝"头脑"——一种滋补早餐。从前老年朋友好这口，如今它也深受年轻人青睐。太原头脑是当地特有的一种风味小吃，已有三百多年的历史，是汤状食品，用料有羊肉、羊髓、酒糟、煨面、藕根、长山药连同黄芪、良姜共计八宗，称为"八珍汤"。太原头脑是由明末清初著名文人、医学家傅山先生想出来的，他在头脑中配了两味草药——黄芪和良姜，这是头脑与其他饮食的不同之处。黄芪味甘性温，能滋补脾胃、活血健肺、延年益寿，对体力虚弱者最有效。

文化剪影 Cultural Outline

Taiyuan, provincial capital of Shanxi, with ancient names as "Jinyang" or "Dragon City". On the beautiful land of Taiyuan, "Soup with Eight Ingredients" is a famous food that has been spread for hundreds of years. It is a **combination**① of health preservation and delicacy, deeply loved by local people.

太原,山西省会,古称"晋阳",也称"龙城"。在秀美的太原大地上,"头脑"是一种已经流传了几百年的名吃,它集养生、美味于一体,深受当地人的喜爱。

Soup with Eight Ingredients has the function of **strengthening**② the body and prolonging life. After a bowl of Soup with Eight Ingredients in winter, you will be sweating and warm all over the body.

太原头脑具有强壮身体、延年益寿的作用。寒冷的冬季,吃一碗太原头脑,你会浑身冒汗、暖意融融。

Fu Shan, inventor of Soup with Eight Ingredients, who saw the people of the Central **Plains**③ were weak and was determined to use their medical skills to help the people strengthen the body and **fight**④ against the **brutal**⑤ rule of the Qing Dynasty.

太原头脑的发明人为傅山,他见中原人民体质较弱,就决心利用自己的医道之长帮助人民强身健体,以抵抗清朝的残酷统治。

 佳句点睛　Punchlines

1. Shanxi people begin their breakfast with a bowl of steaming Soup with Eight Ingredients.

山西人的早餐是从一碗热乎乎的太原头脑开始的。

2. A bowl of Soup with Eight Ingredients is the best breakfast for **defending**⑥ cold.

一碗太原头脑是最好的抗寒早餐。

3. Soup with Eight Ingredients is a famous medicinal food of Shanxi.

太原头脑是山西著名的药膳美食。

 情景对话　Situational Dialogue

A: Are you going anywhere for your **vacation**⑦?

B: Yes, we're making plans for a tour.

A: That will be lovely. Where are you going?

B: We've planned a four-day drive to Taiyuan.

A: Well, you've got to prepare a lot of food.

B: Oh, we'll spend the nights in hotels and enjoy local food as we pass by.

特色美食 第三部分

A: It sounds good. You can have a lot of delicacies.

B: Please tell me about some local delicacies in this city.

A: Ok, what are you interested in?

B: I wanna have some healthy food.

A: What about Soup With Eight Ingredients? It contains different kinds of **nutrients**⑧, with the very local color.

B: Sounds interesting. I'll want a **treat**⑨.

A: 这次你们打算去度假吗?

B: 是的,我们正在制订旅行计划。

A: 太好了,你们要去哪里?

B: 我们计划开车去太原待四天。

A: 那你们得准备很多食物。

B: 我们将在旅馆过夜,顺便品尝一下当地的食物。

A: 听起来不错,你们可以品尝到很多当地美食。

B: 给我分享一些当地美食吧。

A: 好的,你对什么感兴趣?

B: 我想吃一些健康的食物。

A: 太原头脑怎么样?它含有不同种类的营养物质,很有地方特色。

B: 听起来很有趣,我要一饱口福。

饮食文化

生词注解 Notes

① combination /ˌkɒmbɪˈneɪʃən/ n. 结合；组合

② strengthen /ˈstreŋθən/ vt. 强化；加固

③ plains /pleɪnz/ n. 平地；平原

④ fight /faɪt/ vi. 与……打仗；与……斗争

⑤ brutal /ˈbruːtl/ adj. 粗暴的；粗野的

⑥ defend /dɪˈfend/ vt. 防护；为……辩护

⑦ vacation /veɪˈkeɪʃən/ n. 假期；休假

⑧ nutrient /ˈnjuːtrɪənt/ n. 营养素

⑨ treat /triːt/ n. 款待；请客

莜面栲栳栳

Hulls Oats Flour Kaolaolao Noodles

导入语 Lead-in

山西面食不仅品种多样、吃法别致、风格各异，而且可以单独成宴，绝不雷同。在名目繁多的山西面食中，莜面栲栳栳独树一帜，这是山西高寒地区尤其是忻州地区的一种面食小吃，因其形状像"笆斗"而被称为"栲栳"。栲栳是指用竹篾或柳条编制成的一种上下粗细一致的圆筐，形状像斗，是农家专门用来打水或装东西的一种用具。在山西，莜面栲栳栳不仅是一种家常美食，还有犒劳亲朋贵宾之意，人们赋予吃莜面栲栳栳以"牢靠""和睦"等美好象征，每逢老人寿宴、小孩满月或逢节待客，多以此进餐。

 文化剪影　**Cultural Outline**

Hulls Oats Flour, mainly produced in Shanxi, Inner Mongolia, Hebei and other alpine areas, contains **calcium**①, **phosphorus**②, iron, **riboflavin**③ and other nutrients and medicines needed by human body.

莜麦面主产于山西、内蒙古和河北等高寒地区，含有钙、磷、铁、核黄素等多种人体需要的营养元素。

Hulls Oats Flour is a kind of flour with higher nutrition, which can play a very good role in promoting the health of human body. It not only has the function of resisting hunger and cold, protecting the liver and enhancing **immunity**④, but also of strengthening the body, brain and eyes.

莜麦面是一种营养价值比较高的面粉，有利于促进人体健康。莜面不仅有耐饥、抗寒、保肝及增强免疫力的作用，还有强体、健脑、清目的功效。

It is more delicious to have Hulls Oats Flour Kaolaolao Noodles in soup that can be changed with different seasons. In addition to the unique taste, the workmanship of the steamed oat noodle is also very particular.

莜面栲栳栳蘸上汤汁吃更美味，不同季节配制的汤料也不同。除了口感独特之外，做工也很讲究。

佳句点睛 Punchlines

1. The **formation**⑤ of Hulls Oats Flour Kaolaolao Noodles is closely related to the local natural conditions and cultural **customs**⑥.

莜面栲栳栳的形成与当地的自然条件和文化习俗息息相关。

2. The cultural inheritance has an **intimate**⑦ contact with the development of Hulls Oats Flour Kaolaolao Noodles.

文化传承与莜面栲栳栳的发展有着密不可分的联系。

3. Hulls Oats Flour Kaolaolao Noodles is also a kind of cultural **accumulation**⑧.

莜面栲栳栳也是一种文化的积淀。

 情景对话 Situational Dialogue

A: Excuse me, could you tell me where the noodles are in this supermarket?

B: Go the second **aisle**⑨ and you'll find the noodles there.

A: What aisle is that?

B: You'll find it by the oatmeal and breakfast bars.

A: Ok. I understand.

B: Is there anything else you need help?

A: Could you point me where Shanxi Hulls Oats Flour Kaolaolao Noodles is?

B: That's over by the instant noodles.

A: Oh, it is Kaolaolao, which looks wonderful.

B: Alright. That's where it is.

A: Thank you so much for helping me.

B: Let me know if you need anything else.

A: OK.

A: 打扰一下,你能告诉我面条在超市的什么位置吗?

B: 走第二过道就会看到。

A: 那是什么过道?

B: 在燕麦粥和早餐棒旁边。

A: 我知道了。

B: 还有什么需要帮忙的吗?

A: 你能告诉我山西的莜面栲栳栳在哪里吗?

B: 在方便面旁边。

A: 噢,原来这就是栲栳栳,看起来真棒。

B: 对,就在那里。

A: 多谢你帮我。

B: 如果你还需要什么,请告诉我。

A: 行。

生词注解 Notes

① calcium /ˈkælsɪəm/ *n.* 钙

② phosphorus /ˈfɒsfərəs/ *n.* 磷

③ riboflavin /ˌraɪbəˈfleɪvɪn/ *n.* 核黄素

④ immunity /ɪˈmjuːnətɪ/ *n.* 免疫力；豁免权

⑤ formation /fɔːˈmeɪʃn/ *n.* 形成；构成

⑥ custom /ˈkʌstəm/ *n.* 风俗；惯例

⑦ intimate /ˈɪntɪmət/ *adj.* 亲密的；私人的

⑧ accumulation /əˌkjuːmjəˈleɪʃn/ *n.* 积聚；累积

⑨ aisle /aɪl/ *n.* 通道；走道

太谷饼

Taigu Cake

导入语　Lead-in

作为山西人的传统面食小吃,太谷饼因其产于山西晋中市太谷县而得名。太谷饼以其酥软、馨香、色匀、甜而不腻、酥而不碎闻名全国。太谷饼制作精细讲究、色香味俱全,令人回味无穷。太谷饼选用上好的面粉、白糖、食用油和糖稀为原料,糖油配比4比3。外形圆,直径约12厘米,厚约3厘米,边心厚薄均匀,表皮茶黄,粘有脱皮芝麻仁。太谷饼既能当茶点,又能旅行时食用,既是晋商饮食文化的象征,又是糕点王,享有"平遥的牛肉,太谷的饼"之美誉。2007年被列入《中国国家地理标志产品保护目录》。

文化剪影 Cultural Outline

Taigu Cake is a typical representative of the food culture of Shanxi **merchants**①, which is a **folk**② snack with cultural foundation with the development of history and culture.

太谷饼是晋商饮食文化的典型代表,伴随着历史与文化的发展而产生,是一道颇具文化气息的民间小吃。

Taigu Cake, with a history of more than 150 years, is a famous traditional snack in Shanxi Province and enjoys the fame of "The King of Pastry."

太谷饼迄今已有一百五十余年历史,是誉满山西的传统小吃,享有"糕点之王"的美誉。

Shanxi **occupies**③ an important position in the process of human food civilization, among which Taigu Cake is one of the intangible cultural heritages; the outside is **crisp**④, the inside tender.

山西在人类饮食文明进程中占有重要一席,太谷饼是山西的非物质文化遗产之一,其外皮酥香,内里绵软。

佳句点睛 Punchlines

1. Taigu Cake is a perfect combination of traditional craft and

modern civilization.

太谷饼是传统工艺与现代文明的完美结合。

2. The difference between Taigu Cake and other pastry food is that there're no **additives**⑤ and **preservatives**⑥ in the cake.

太谷饼和其他糕点不同的是，饼里不含任何添加剂和防腐剂。

3. The production process of Taigu Cake is strict and materials are also very **fastidious**⑦.

太谷饼的制作工艺严格，用料也十分考究。

 ## 情景对话　Situational Dialogue

A: I think I'm going to the supermarket today.

B: Do we need food?

A: Yeah, I think so.

B: What are you gonna get?

A: I'm not sure what we need.

B: Maybe you should go and look in the **refrigerator**⑧.

A: Could you do it for me, and make a list of things that we need?

B: Just get the basics.

A: Like what?

B: You know. Get some eggs, milk and bread. Right, don't forget to bring some Taigu Cake, it's so yummy.

A: Just go and make a list for me, please.

B: Fine, I'll go to do it.

A: 我今天要去超市。

B: 我们需要买食物吗?

A: 是的,我想是的。

B: 你要买什么?

A: 我不确定我们需要什么。

B: 你可以去看看冰箱。

A: 你能去看一下吗? 然后把我们需要的东西列个清单。

B: 买基本款就可以了。

A: 比如买什么?

B: 你知道的,买一些鸡蛋、牛奶和面包。对了,别忘了买点儿太谷饼,它太好吃了。

A: 去帮我列个单子吧。

B: 好的,我这就去。

生词注解 Notes

① merchant /ˈmɜːtʃənt/   *n.* 商人

② folk /fəʊk/   *adj.* 民间的

③ occupy /ˈɒkjupaɪ/   *vt.* 占用;占有

④ crisp /krɪsp/   *adj.* 脆的;新鲜的

⑤ additive /ˈædətɪv/　　n. 添加剂；食物添加剂
⑥ preservative /prɪˈzɜːvətɪv/　　n. 防腐剂；保存剂
⑦ fastidious /fæˈstɪdɪəs/　　adj. 挑剔的；苛求的
⑧ refrigerator /rɪˈfrɪdʒəreɪtə(r)/　　n. 冰箱；冷藏库

兰州牛肉拉面

Lanzhou Beef Ramen

 导入语 Lead-in

兰州牛肉拉面别名"牛大面",又称为"兰州清汤牛肉拉面",起源于清朝嘉庆年间,从河南怀庆府清化镇(今河南博爱县)传到了甘肃兰州,并在那里发扬光大,以"汤镜者清、肉烂者香、面细者精"见长,达到了"闻香下马,知味停车"的妙境,被誉为"中华第一面"。小小一碗面已经从家庭作坊发展成为香飘全国的一道美食。这碗面不仅可以充饥,而且拥有一份独特情怀,它蕴含着兰州味道,承载着文化底蕴。

饮食文化

文化剪影 Cultural Outline

Lanzhou Beef Ramen is a famous traditional Chinese food, in which white **radish**①, chilli oil, **coriander**② and yellow noodles in clear beef broth are teemed with the color, flavor and taste.

兰州牛肉拉面是中国的一种传统名食,白萝卜、红辣椒油、香菜和黄面条煨在牛肉清汤里,色香味俱全。

Chinese people like to eat Lanzhou Beef Ramen not only because it is a **convenient**③ hot fast food conforming to the Chinese taste, but because of **diligence**④ and simplicity contained in the noodles, which complies with the heart of the Chinese people.

中国人喜欢吃兰州牛肉拉面,不仅因为它是适合中国人口味的一道便利的热食快餐,而且因为面中蕴含的勤劳和质朴,顺应了中国人的心意。

The establishment of relevant industry standards and the modern **enterprise**⑤ operation help Lanzhou Beef Ramen move towards the wholly new fast food mode and **brand**⑥ on the way to the development.

相关行业标准的制定以及现代化的企业运营,帮助兰州牛肉拉面走向全新的快餐发展模式,迈向品牌化发展道路。

佳句点睛 Punchlines

1. Lanzhou Beef Ramen is changing constantly with the development of the times.

随着时代的发展,兰州牛肉拉面也在不断发生变化。

2. The process of making Lanzhou Beef Ramen is **amazing**⑦ to foreigners.

兰州牛肉拉面的制作过程让外国人感到奇妙无比。

3. Only when the soup of Lanzhou Beef Ramen is authentic can the original flavor be guaranteed.

只有正宗的汤料才能确保做出原汁原味的兰州牛肉拉面。

情景对话 Situational Dialogue

A: We come at the worst time, for there're so many people here.

B: That's true, I'm starving. Let's just have some noodles, alright?

A: Let's try Lanzhou Beef Ramen.

B: Hey look! There're two seats here.

A: Hand-pulled noodles with beef are really popular in China. We can see lots of similar restaurants along the road.

B: Yes, it's praised as one of the three Chinese fast food by the

饮食文化

Chinese Cuisine **Association**®.

A: There's a great variety of Chinese noodles, which vary according to their regions of production, ingredients, shapes and manners of preparation.

B: Absolutely right.

A: Would you like anything to drink?

B: How about orange juice?

A: OK, let's order two bowls of Ramen and two bottles of orange juice.

B: OK.

A: 我们来的不是时候,这里太多人了。

B: 是啊,我饿死了。我们简单吃点儿面吧,好吗?

A: 我们尝尝兰州牛肉拉面吧。

B: 看!这里有两个座位。

A: 手工牛肉拉面在中国的确很受欢迎。我们一路上可以看到很多这样的餐馆。

B: 是的,它被中国烹饪协会誉为三大中式快餐之一。

A: 中国的面条种类繁多,根据其产地、配料、形状和制作方法的不同而有所区别。

B: 没错。

A: 你想喝点儿什么吗?

B: 橙汁怎么样?

A: 好的,我们要两碗拉面和两瓶橙汁。

B: 好嘞。

生词注解 Notes

① radish /ˈrædɪʃ/ n. 萝卜；小萝卜

② coriander /ˌkɒrɪˈændə(r)/ n. 芫荽；香菜

③ convenient /kənˈviːnɪənt/ adj. 方便的；适当的

④ diligence /ˈdɪlɪdʒəns/ n. 勤奋；勤勉

⑤ enterprise /ˈentəpraɪz/ n. 企业；进取心

⑥ brand /brænd/ n. 品牌

⑦ amazing /əˈmeɪzɪŋ/ adj. 令人惊异的

⑧ association /əˌsəʊsɪˈeɪʃn/ n. 协会；联盟

热干面

Hot-dry Noodles

导入语 Lead-in

武汉饮食融汇各地特色，在众多小吃中，热干面最有名气，也是当地人"过早"（吃早餐）的首选小吃。热干面作为湖北省的特色小吃，与北京炸酱面、河南烩面、兰州牛肉拉面、山西刀削面和四川担担面等同称为"中国十大名面"。

热干面的面条纤细爽滑，根根筋道，色泽黄润，滋味鲜美。面条事先煮熟，过冷和过油后，浇上醋、辣椒油，拌以香油、芝麻酱、五香酱等，可谓色香味俱全。热干面可以干吃，也可以与其他食材一起加汤吃，既可以素吃，也可以荤吃。热干面就像武汉这座城市的个性一样兼容并蓄。

文化剪影 Cultural Outline

China has a rich history of pasta, bringing together a variety of cultural **ingredients**①. Wuhan Hot-dry Noodles, attractive in color, smell and taste, is a model of the blending of history and multi-cultures.

中国面食历史丰富,汇集了多种文化成分。色香味俱全的武汉热干面是历史与多元文化交融的典范。

Hot-dry Noodles can represent the character of Wuhan people, whose appearance is rough, but the core is delicate. The **inclusive**② and **delicate**③ character has become an **indispensable**④ part of the local cuisine culture as well.

武汉热干面代表着武汉人的性格,外表粗犷,但内核精致。兼容并蓄、细腻粗犷也成为当地饮食文化中不可缺少的一部分。

It's the Hot-dry Noodles that make people linger on Wuhan for quite a long time, because it represents Wuhan people's passion for life.

正是热干面使人对武汉久久无法忘怀,因为它代表着武汉人对生活的热情。

 佳句点睛 Punchlines

1. Pastry is closely related to people's daily life.

面点与人们的日常生活息息相关。

2. Chinese pastry has different styles due to different regional **distinctions**⑤.

中国面点因不同的区域差异而形成不同的风格特色。

3. Hot-dry Noodles **integrates**⑥ the characteristics of northern and southern cuisines in its development.

热干面在发展中兼具南北美食特色。

 情景对话　**Situational Dialogue**

A: Where did you go this summer vacation?

B: My elder sister was gonna marry in Wuhan, so we went to Wuhan for a visit.

A: What **impressed**⑦ you most about Wuhan?

B: Hot-dry Noodles. We are used to haveing **porridge**⑧ for breakfast, but Wuhan people eat noodles in the morning.

A: Don't we have noodles for lunch?

B: Yeah, but if you go to Wuhan for breakfast, almost everyone walks with a bowl of Hot-dry Noodles.

A: Walking with a bowl of Hot-dry Noodles?

B: Yes, they seem like to eat Hot-dry Noodles while walking.

A: People in Wuhan are crazy about Hot-dry Noodles. It's hard for foreigner to understand.

B: Maybe it's because local people are used to the taste of it.

A: When I have a chance to go to Wuhan, I must try the authentic Hot-dry Noodles.

B: It has become the symbol of Wuhan.

A: 这个暑假你去哪里了?

B: 我姐要在武汉结婚,我们顺便去武汉转了一圈。

A: 武汉给你留下最深刻印象的是什么?

B: 热干面,我们习惯早上喝粥,但武汉人早上就吃面。

A: 我们不是中午才吃面吗?

B: 是啊,但如果你去武汉吃早饭,就会发现走在你身边的人手上几乎都端着一碗热干面。

A: 一边走一边端着吃热干面?

B: 是的,他们似乎很喜欢端着热干面边走边吃。

A: 武汉人对热干面的痴迷,外地人很难理解。

B: 可能是当地人已经习惯了热干面的味道。

A: 有机会去武汉,一定要尝尝正宗的热干面。

B: 热干面已然成为武汉的符号。

生词注解 Notes

① ingredient /ɪnˈgriːdɪənt/ n. 原料;要素

② inclusive /ɪnˈkluːsɪv/ adj. 包括的;包含的

③ delicate /ˈdelɪkət/ *adj.* 精美的；雅致的

④ indispensable /ˌɪndɪˈspensəbl/ *adj.* 不可缺少的；绝对必要的

⑤ distinction /dɪˈstɪŋkʃn/ *n.* 特性；差别

⑥ integrate /ˈɪntɪɡreɪt/ *vt.* 使……结合；使……一体化

⑦ impress /ɪmˈpres/ *vt.* 给……留下深刻印象

⑧ porridge /ˈpɒrɪdʒ/ *n.* 粥；麦片粥

精武鸭脖

Jingwu Duck-neck

 导入语 Lead-in

湖北是中华民族和中国古代文化的发祥地之一。光辉灿烂的楚文化不仅让湖北享誉海内外，也使湖北的饮食文化独具特色。从武昌鱼、豆皮、热干面、糊汤粉到精武鸭脖，无不叙说着湖北悠久的美食文化。正宗的精武鸭脖制作十分考究，须用上等辣椒，外加五香、花椒、八角、陈皮等三十八种香辛料熬煮近一个小时作为卤汁，再将鸭脖放入卤水中煮近两个小时，才算大功告成。由于核心配料不大相同，因此各家鸭脖的口味又有所区别。这种风味独特、别具一格的卤鸭脖成了武汉的特色美食，也成了武汉的一张靓丽名片。

 文化剪影 **Cultural Outline**

Jingwu Duck-neck is one of the most famous traditional snacks in Wuhan of Hubei Province, which got its name from Jingwu Road in Hankou. Today, Jingwu Road is well known for a small duck-neck.

精武鸭脖是湖北武汉最有名的传统小吃之一，因起源于汉口的精武路而得名。时至今日，精武路因一根小小的鸭脖而声名远扬。

Strolling in the narrow streets, you are **overwhelmed**① with a burst of spicy **enticing**② smell, mouth-watering. Looking ahead, lanes on both sides are full of specialty shops and stalls, and the phrase of "Jingwu Duck-neck" can be seen everywhere.

漫步于窄窄的小街上，一阵阵香辣诱人的味儿扑鼻而来，令人垂涎。放眼望去，小巷两边满是专卖店和大排档，"精武鸭脖"的字样随处可见。

The duck-neck is a new contribution of Wuhan people to Chinese cuisine. The people of Wuhan would like to eat duck-necks because it tastes strong enough. The duck-neck is yummy because the secret is all in the soup; the soup is freshly fragrant because of the finest **spices**③.

鸭脖是武汉人对中华美食的新贡献。武汉人喜欢吃鸭脖，因为它味足够劲。鸭脖好吃的秘诀全在汤料里；而汤料鲜香源于上等的香料。

佳句点睛 Punchlines

1. The duck-neck is **lusciously**④ **delish**⑤ and a leisure food for the old and young.

鸭脖香鲜美味,是老少皆宜的休闲食品。

2. A small duck-neck has changed the fate of countless common people.

一根小小的鸭脖改变了无数草根的命运。

3. Jingwu Duck-neck can be seen everywhere across the country and has formed some new schools.

精武鸭脖在全国各地随处可见,并形成了一些新流派。

情景对话 Situational Dialogue

A: Hey, what are you doing now?

B: I'm watching TV. It's really wonderful.

A: What's on?

B: It's the World Cup that I have been waiting for four years.

A: I can't imagine why so many people would like to see the ball kicked here and there.

B: Because we can have a lot of fun.

A: Oh, I guess you will stay up to watch it.

B: Yes, of course. It's a good opportunity to see my favorite stars. Is there anything in the **fridge**⑥ I can eat?

A: There's some fruit for you.

B: Oh, I really wanna have Jingwu Duck-neck. Would you like to buy some for me, Mommy?

A: OK, but you can't shout loudly at midnight.

B: I promise.

A: 嘿,你在干什么呢?

B: 我在看电视,真的很棒。

A: 是什么节目?

B: 世界杯,我已经等了四年了。

A: 我无法想象为什么那么多人喜欢看足球被踢来踢去。

B: 因为我们可以得到很多乐趣。

A: 噢,我猜你会熬夜看的。

B: 当然,这是看我最喜欢的球星的好机会。冰箱里有什么吃的东西吗?

A: 有些水果。

B: 我真想吃精武鸭脖。妈妈,你能给我买一些吗?

A: 好的,但你不能半夜大声喊叫。

B: 我保证。

生词注解 Notes

① overwhelm /ˌəʊvəˈwelm/ *vt.* 压倒；覆盖

② enticing /ɪnˈtaɪsɪŋ/ *adj.* 诱人的；有吸引力的

③ spice /spaɪs/ *n.* 香味料；调味料

④ lusciously /ˈlʌʃəslɪ/ *adj.* 甘美的；味香的

⑤ delish /ˈdelɪ/ *adj.* 美味的；可口的

⑥ fridge /frɪdʒ/ *n.* 电冰箱

南翔小笼包

Nanxiang Small Steamed Buns

导入语 Lead-in

南翔小笼包诞生于上海市嘉定区南翔镇日华轩点心店,曾被称为"南翔大肉馒头""古猗园小笼",是上海家喻户晓的美味点心,其制作技艺也是上海市首批非物质文化遗产。南翔小笼包源自清朝同治年间,以不发酵的精面粉为皮,馅料用猪腿精肉手工剁成,肉馅里再加上肉皮冻,以皮薄、馅丰、肉嫩汁多、味鲜和形美著称,出笼时呈半透明状,小巧玲珑。咬一口南翔小笼包,肉馅里鲜美的汤汁令人啧啧称赞、回味无穷,是千年古镇的一张独特的文化名片。如今,南翔小笼包远销美国、加拿大、澳大利亚、英国等国家。

文化剪影 Cultural Outline

Nanxiang Small Steamed Buns originated from the ancient town of Nanxiang, which has a long history, profound cultural heritage and prosperous economy. The **exquisite**① steamed buns can meet the need of local people as they pursue the art of eating.

南翔小笼包起源于古镇南翔,南翔历史源远流长,文化底蕴深厚,经济繁荣。当地人追求饮食艺术,而精美的小笼包满足了他们的美食需求。

With the efforts of generations of **successors**②, the fame of Nanxiang Small Steamed Buns is rising day by day; nowadays it has formed a unique production technology and secret **recipe**③ and eventually become the **time-honored**④ brand.

经历了几代传人的努力,南翔小笼包的名声与日俱增,如今已经形成了独特的制作工艺和配制秘方,终成老字号。

The unique charm of Nanxiang Small Steamed Buns lies in exquisite **workmanship**⑤ and meat stuffing. The eating method can be steamed, fried or boiled. It's a temptation both in sight and smell.

南翔小笼包的独特魅力在于它的做工讲究,肉馅精良。南翔小笼包的食用方法可以是"蒸""炸",也可以是"烧"。无论在视觉还是在嗅觉上都十分诱人。

 佳句点睛 **Punchlines**

1. Baozi makes more people in the world know the long-standing Chinese cuisine culture.

包子让世界上的更多人了解到源远流长的中华美食文化。

2. Nanxiang Small Steamed Buns has evolved to satisfy the different tastes of local people.

南翔小笼包不断改良以满足各地方人们的不同口味。

3. Nanxiang Small Steamed Buns is also **infused**⑥ the concept of modern culture.

南翔小笼包也注入了现代文化的理念。

 情景对话 **Situational Dialogue**

A: Welcome to listen to our program — *This is Shanghai*!

B: Xiaosong, what would you like to introduce today?

A: Today I'm gonna talk about a famous cuisine in Shanghai.

B: As a modern city, Shanghai is a shining pearl inlaid on the **oriental**⑦ land. I would like to introduce Nanxiang Small Steamed Buns with a history of over one hundred years in today's program. Nanxiang Small Steamed Buns is my father's favorite. He often goes to

Town God Temple (Chenghuang Temple) to eat Nanxiang Small Steamed Buns.

A: It seems that Nanxiang Small Steamed Buns is very popular.

B: I like it, too. Compared with other steamed buns, Nanxiang Small Steamed Buns' secret lies in the juice inside the bun.

A: After sucking out the juice from the bun with a straw, the taste would be **lingering**⑧.

B: People would also like to have a bowl of wonton together with steamed buns.

A: As far as I know, every procedure of Nanxiang Small Steamed Buns has clear standards.

B: It is really one of delicacies that we must try in Shanghai.

A: 欢迎各位收听今天的节目——《这就是上海》！

B: 小松，今天你要给大家介绍什么呢？

A: 今天我会跟大家讲一道上海著名美食。

B: 上海作为现代都市，是一颗镶嵌在东方大地上的璀璨明珠。在今天的节目中，我来讲一讲上海具有百年历史的南翔小笼包。南翔小笼包是我爸的最爱。他经常跑到城隍庙去吃南翔小笼包。

A: 看来南翔小笼包依然很受欢迎啊！

B: 我也很喜欢。与其他包子相比，南翔小笼包的秘诀在于包子里的汤汁。

A: 用吸管吸出汤包中的汤汁后，令人回味无穷。

B: 吃包子时人们还喜欢配上一碗馄饨。

饮食文化

A: 据我所知,南翔小笼包的每一道工序都有明确标准。

B: 南翔小笼包真是我们来上海必须品尝的美食之一。

生词注解　Notes

① exquisite /ɪkˈskwɪzɪt/　*adj.* 精致的;细腻的

② successor /səkˈsesə/　*n.* 继承人

③ recipe /ˈresəpɪ/　*n.* 食谱;烹饪法

④ time-honored /ˈtaɪmɒnəd/　*adj.* 由来已久的;因为古老而受敬仰的

⑤ workmanship /ˈwɜːkmənʃɪp/　*n.* 手艺;工艺

⑥ infuse /ɪnˈfjuːz/　*vt.* 灌输;使……浸泡

⑦ oriental /ɔːrɪˈentl/　*adj.* 东方的;东方人的

⑧ lingering /ˈlɪŋɡərɪŋ/　*adj.* 绵长的;久久不散的

生煎包

Pan-fried Pork Buns

 导入语　Lead-in

　　生煎包是上海、浙江、江苏、广东等地的特色传统小吃，最早出现在一百多年前的上海茶馆里。后来，人们不品茶时，也想以"茶点"作为快餐来代替正餐。为了顺应人们的饮食需求，生煎包也逐渐走出茶馆变为人们的早点，成为街头小吃。"生煎包"也叫"生煎馒头"，生煎包的制作过程并不复杂，首先用半发酵面团，包上适量肉馅，下油锅煎底，加水、焖锅，待至水干、油分被充分吸收即可。其特点是面皮油润柔软，包底金黄脆香，馅心鲜嫩适口。生煎包营养丰富，具有驱寒、健脾、养胃、顺气、中和的效果。

 文化剪影　Cultural Outline

Pan-fried Pork Buns is a traditional **dim sum**① in Shanghai, with a history of more than one hundred years. It seems simple, but each step is particular from **kneading**② the dough to duration and degree of heating.

生煎包是上海传统点心,已有上百年的历史。生煎包看似简单,但从和面到火候的把握,每一步都很讲究。

Pan-fried Pork Buns is a special delicious snack in Shanghai. Its skin is **crisp**③ and the stuffing is juicy and yummy. It **represents**④ the taste of the city of Shanghai.

生煎包是一种上海特有的美味小吃,其皮脆汁多,鲜美无比。它代表着上海这座城市的味道。

The soul of Pan-fried Pork is soup, which is made by boiling ingredients rich in **collagen**⑤ such as pig skin, pig feet and chicken claws. After cooling and **solidifying**⑥, the soup will be **frozen**⑦, cut it up and mixed it with the stuffing.

生煎包的灵魂是汤汁,是将猪皮、猪蹄、鸡爪等富含胶原蛋白的食材煮成高汤,冷却凝固成冻状,然后将汤冻切碎、拌入肉馅中制成。

佳句点睛 **Punchlines**

1. Pan-fried Pork Buns is not only a kind of food, but it is also full of rich connotation.

生煎包不仅是一种食品,还有着丰富的内涵。

2. Pastry is an important part of Chinese cuisine and an indispensable food in people's daily life.

面点是中国烹饪的重要组成部分,是人们餐桌上不可或缺的食物。

3. With the improvement of making techniques, Pan-fried Baoz Pork Buns has gradually formed different schools.

随着制作技艺的提高,生煎包逐渐形成了不同流派。

 情景对话 **Situational Dialogue**

A: What should we have for dinner this evening?

B: Are you asking me?

A: Yes, I am. I really don't feel much like cooking, but the family must eat.

B: What about eating outside?

A: I don't think the food outside is healthy, especially the food in summer.

B: OK, it's up to you.

A: Can you give me a hand in the kitchen? I don't think I can finish everything by myself.

B: What do you want me to do?

A: I need you to chop the pork for me. Today we are doing to have Pan-fried Pork Buns.

B: What? Do you know how to do it?

A: I am not sure. But I have a recipe. I can do it by the **recipe**[8].

B: You'd better ask Lin for help. She is good at cooking.

A: Alright.

A: 今晚吃什么?

B: 你是在问我吗?

A: 是啊。我实在不想做饭,但全家总得吃饭吧。

B: 去外面吃怎么样?

A: 我觉得外面的食物不健康,尤其是在夏天。

B: 好吧,你来决定。

A: 你能来厨房帮我一下吗? 我一个人做不完所有的事情。

B: 你想让我做什么?

A: 我需要你帮我剁猪肉,今天我们要做生煎包。

B: 什么? 你知道怎么做吗?

A: 我不确定。不过,我有个食谱,可以按食谱做。

B: 你最好找林帮忙。她擅长烹饪。

A: 好的。

生词注解 Notes

① dim sum /ˌdɪm ˈsʌm/ n. 点心

② knead /niːd/ vt. 揉合；揉捏

③ crisp /krɪsp/ adj. 脆的；新鲜的

④ represent /ˌreprɪˈzent/ vt. 代表；表现

⑤ collagen /ˈkɒlədʒən/ n. 胶原蛋白

⑥ solidify /səˈlɪdɪfaɪ/ vt. 使……凝固；使……固化

⑦ freeze /friːz/ vt. 凝固；结冰

⑧ recipe /ˈresəpɪ/ n. 食谱；烹饪法

饮食文化

蟹壳黄烧饼

Xiekehuang Sesame Seed Cake

 导入语 Lead-in

江南传统名小吃蟹壳黄烧饼因饼形似蟹壳,烤熟后色泽如蟹背一样深红而得名,也称为"小炉饼""小烧饼"。它采用油酥面加酵面制坯,做成扁圆型饼。饼面粘上一层芝麻,贴在炉壁上经烘制而成。馅料有咸有甜,咸的有葱油、鲜肉、蟹粉、虾仁等,甜的有白糖、玫瑰、豆沙、枣泥等。"未见饼家先闻香,入口酥皮纷纷下";陶行知先生有诗赞曰:"三个蟹壳黄,两碗绿豆粥,吃到肚子里,同享无量福。"蟹壳黄烤出来能否酥脆、层层剥落,都取决于面皮,做皮的面粉要好,要舍得放菜籽油,生面团须揉得铿亮,这样烤出来才好吃。

文化剪影　Cultural Outline

Xiekehuang Sesame Seed Cake is actually a kind of miniature **sesame**[①] pie, as we know the pie in the north often serves as the staple food while the pie in the south is exquisite, which is a dessert at most.

蟹壳黄烧饼其实就是一种迷你型的芝麻烧饼,北方的烧饼常充作果腹的主食,而南方的烧饼则讲究精巧,至多是点心。

The diets of the north and south of China vary greatly, but Xiekehuang Sesame Seed Cake offers a string of the connections between the regions.

中国南北方饮食差异巨大,但蟹壳黄烧饼体现了各地饮食之间千丝万缕的联系。

Xiekehuang is so named because of its yellow shape like a crab shell, which is a **dough**[②] made of **pastry**[③] and yeast, made of small oval shaped pie, outside with a layer of sesame, baking in the oven wall.

蟹壳黄因形似蟹壳而得名,是用酥皮和酵母做成面团,外面有一层芝麻,在炉壁上烘烤而成。

饮食文化

 佳句点睛　Punchlines

1. Xiekehuang Sesame Seed Cake can be found around the town in any season.

在镇上，任何季节都能找到蟹壳黄。

2. The production of Xiekehuang was paid great attention to the selection of materials.

蟹壳黄的制作特别重视用料的选择。

3. From the west of China to the east, the fire-baked pasta changes into different appearances.

从中国西部到东部，这种火烤的面食变化出不同的模样。

 情景对话　Situational Dialogue

A: Could you talk about how Chinese cuisine culture impresses you generally?

B: Well, cooking has occupied a lofty position in Chinese culture throughout history. The great Chinese **philosopher**① Lao Tzu once stated that "governing a great nation is much like cooking a small fish".

A: Do you like Chinese food and what is your favorite?

B: Yes, I do. I love Chinese food such as soybean milk, Xiekehuang and, so on.

A: Xiekehuang?

B: It is the sesame seed cake, which is named because its yellow shape is like a crab shell.

A: Do you think Chinese cooking is healthy?

B: Yes. Color, aroma, and flavor are not the only principles to be followed in Chinese cooking; nutrition is also an important concern. The principle of the harmonization of foods can be traced back to Yi Yin, the Shang Dynasty scholar.

A: Can you explain?

B: He related the five flavors of sweetness, sour bitterness, piquancy, and saltiness to the nutritional needs of the five vitals of the body — the heart, livers, **spleen**⑤/**pancreas**⑥, lungs, and kidneys, stressing on their role in maintaining health.

A: You certainly know more about Chinese food than many Chinese.

B: Because I love Chinese food! In fact, many of the plants used in Chinese cooking, such as **scallions**⑦, fresh ginger root, garlic, tree fungus, and so forth, have properties of preventing and **alleviating**⑧ various illnesses.

A: What interests you most in your studies of China's food culture?

B: I think the Chinese food **symbolism**⑨ is most interesting. Noodles are a symbol of longevity, chicken forms part of the symbolism of the dragon and **phoenix**⑩, and the seeds represent bearing many children in

饮食文化

Chinese culture.

A: 你能不能谈一下对中国饮食文化的整体印象？

B: 一直以来，饮食在中国文化中都占据重要地位。中国哲学家老子曾经说过："治大国若烹小鲜。"

A: 你喜欢中餐吗？你最爱吃的中国食品是什么？

B: 是的，我非常喜欢吃中餐，如豆浆、蟹壳黄等。

A: 蟹壳黄？

B: 它是芝麻烧饼，因形似蟹壳而得名.

A: 你认为中餐有益健康吗？

B: 是的。色、香、味并不是评判一道菜肴的全部标准，菜肴的营养价值在中餐中也很重要。食物中均衡和谐的概念可以追溯到商朝的学者伊尹。

A: 你能解释一下吗？

B: 伊尹解释了五味与五脏的关系，以及五味对健康的作用。

A: 你对中餐的了解比很多中国人都透彻。

B: 那是因为我非常喜欢中餐！实际上很多中餐烹饪中使用的植物，如葱、姜、蒜、木耳等，都有防治疾病、减轻病痛的作用。

A: 在你研究中国饮食文化的过程中，觉得哪部分最有意思？

B: 我觉得中国食物的象征意义最有意思。面条象征长寿，鸡肉是龙凤呈祥里的"凤"，瓜子在中国文化里象征着多子多孙。

生词注解 Notes

① sesame /ˈsesəmɪ/ n. 芝麻；通行证

② dough /dəʊ/ n. 生面团；金钱

③ pastry /ˈpeɪstrɪ/ n. 油酥点心；面粉糕饼

④ philosopher /fəˈlɒsəfə(r)/ n. 哲学家；哲人

⑤ spleen /spliːn/ n. 脾脏

⑥ pancreas /ˈpæŋkrɪəs/ n. 胰腺；胰脏

⑦ scallion /ˈskælɪən/ n. 葱；青葱

⑧ alleviate /əˈliːvɪeɪt/ vt. 缓和；减轻

⑨ symbolism /ˈsɪmbəlɪzəm/ n. 象征意义；符号论

⑩ phoenix /ˈfiːnɪks/ n. 凤凰；死而复生的人

长沙臭豆腐

Changsha Preserved Smelly Tofu

导入语　Lead-in

湖南长沙传统特色名小吃臭豆腐应该是中国最出名的臭味美食了，长沙当地人称之为"臭干子"。臭豆腐颜色墨黑、香辣、焦脆、细嫩。臭豆腐是一种散发浓烈气味的发酵豆腐，分臭豆腐干和臭豆腐乳两种，是由豆腐浸泡在蔬菜发酵的盐水中制成，主要食材有黄豆、辣椒、豆豉，调料有酱油、卤水、盐、熟石膏。首先制出卤水，经过几周甚至几个月的发酵，逐渐形成一种发酵液体，之后再加入豆腐进一步发酵。臭豆腐通常是油炸的，加以豆瓣酱或甜酱，使这种小吃闻起来臭，吃起来香。古医书记载，臭豆腐可以寒中益气，和脾胃，清热散血，下大肠浊气。

特色美食 第三部分

文化剪影 Cultural Outline

The Chinese have unlimited **enthusiasm**① and genius for eating, including smelly food. It's **microbe**② that works, no matter mass produced fried fermented tofu or "**stale**"③ food at home.

中国人对于吃有着无限热情与天赋,其中臭味美食也在其列。无论是批量生产的臭豆腐还是家中"变味变霉"的食物,都是微生物在发挥作用。

Tofu has high **nutrition**④ value and is favored by the healthy pursuers. As a representative of tofu food, the fried fermented tofu gives off an **indescribable**⑤ smell when it **ferments**⑥. Some people will stay away from it, while others will love it deeply.

豆腐因其营养价值高而受到健康追求者的追捧。作为豆腐中的经典代表,臭豆腐经发酵后会散发出一种难以描述的气味,有人敬而远之,有人却情有独钟。

There're various production methods for tofu, while Changsha Preserved Smelly Tofu is unique. The black appearance has unique characteristic, the smell is stinky at first, but the taste is yummy.

豆腐的做法多样,其中长沙臭豆腐独具特色。黑黑的外表闻起来臭气扑鼻,吃起来却别有一番风味。

佳句点睛 Punchlines

1. Tofu is the representative of Chinese wisdom, which is also the treasure for **vegetarians**⑦.

豆腐是中国人智慧的结晶,是素食者的珍宝。

2. Tofu culture is **spreading**⑧ to every corner of the world.

豆腐文化正深入到世界的各个角落。

3. Changsha Preserved Smelly Tofu is a unique snack in China.

长沙臭豆腐是中国小吃一绝。

情景对话 Situational Dialogue

A: What does it smell like? Do you smell it?

B: It's the smell of Changsha Preserved Smelly Tofu.

A: What a special smell!

B: Yes, the smell is not very good, but it tastes good.

A: Mom, you said we couldn't eat bad food. How can we eat when it stinks?

B: There're a lot of foods, such as preserved eggs, which go bad but they're still edible.

A: Is that good for our health?

B: We'd better have in moderation, but not in excess.

A: Can I have a little sugar next time?

B: No, you have **decayed**⑨ teeth, you can't eat sugar.

A: Oh, you've made me stir-fried tofu and tofu soup, but I've never eaten Changsha Preserved Smelly Tofu. I wanna have a try.

B: OK, let's buy it together.

A: 这是什么味儿啊？你闻到了吗？

B: 长沙臭豆腐的味道。

A: 这味道好特别啊!

B: 是啊,闻起来臭,吃起来香。

A: 妈,你说坏的东西不能吃,可它都臭了怎么还吃呢？

B: 我们有很多食物,比如皮蛋,虽然变质了,但仍然可以吃。

A: 那对我们身体好吗？

B: 可以适量吃,但不能多吃。

A: 下次我稍微吃点儿糖可以吗？

B: 不行,你都有蛀牙了,不能吃糖。

A: 好吧,你给我做过炒豆腐、豆腐汤,可我还没吃过长沙臭豆腐。我想尝尝。

B: 好,我们一起去买吧。

 生词注解 Notes

① enthusiasm /ɪnˈθjuːzɪæzəm/ *n.* 热忱；热情

② microbe /ˈmaɪkrəʊb/ n. 细菌；微生物

③ stale /steɪl/ adj. 不新鲜的；陈腐的

④ nutrition /njuˈtrɪʃn/ n. 营养；营养品

⑤ indescribable /ˌɪndɪˈskraɪbəbl/ adj. 难以形容的；不可名状的

⑥ ferment /fəˈment/ vt. 使……发酵；激起

⑦ vegetarian /ˌvedʒəˈteərɪən/ n. 素食者；素食主义者

⑧ spread /spred/ vi. 遍及；弥漫

⑨ decayed /dɪˈkeɪd/ adj. 腐烂的；腐败的

口味虾

Spicy Crawfish

导入语 Lead-in

口味虾,又名"麻辣小龙虾""长沙口味虾"和"香辣小龙虾"等,是湖南省的传统名小吃,以小龙虾配干红辣椒、植物油、精盐、味精、酱油、白醋、料酒、生姜、葱花、香菜末精制而成,色泽红亮,滑嫩、香辣、鲜浓。口味虾在北方称为"小龙虾",最初生长在一些水塘里,人们很少把它当成桌上的菜肴。后来,经过很多厨艺高手不断探讨它的烹制方法,时至今日,尤其是在夏季的长沙,吃口味虾早已成了一道不可缺少的风景线。口味虾的独到之处在于它的奇辣无比,而不同的口味虾店都有各自独特的烹制方法,这样才会吸引更多的"虾客"。

文化剪影 Cultural Outline

Changsha is famous for its own culture since ancient times, with a profound cultural origin and unique diet features. Therefore, both foreign diners and native people are **immersed**① in the delicacies.

长沙自古以湘楚文化闻名,具有深厚的文化渊源和独特的饮食特色。因此,无论是外地食客还是本地人,都沉醉于美食之中。

Hunan is extremely rich in products and local specialties. While tasting deliacies, we should also take care of nutrition and health.

湖南的物产和地方特色美食都极为丰富。在品尝美食的同时,我们也要关注营养与健康。

Unlike other delicacies, the spicy **crayfish**② breaks down regional **discrimination**③ and taste **barriers**④ in a flash of lightning to capture the taste buds of all Chinese people swiftly.

没有一种美食像口味虾一样以迅雷不及掩耳之势打破地域歧视和口味壁垒,迅速掳获所有国人的味蕾。

佳句点睛 Punchlines

1. The crayfish will always **dominate**⑤ the night table in the city as long as the summer is not over.

只要夏日未尽,小龙虾便永远在城市的宵夜中占有一席之地。

2. Changsha is, so to speak, a promised land for both tourists and foodies.

可以说,长沙既是游人的福地,也是吃货的乐园。

3. Hunan people can find the joy and satisfaction of life from delicacies.

湖南人能从美食中找到生活的快乐和满足感。

情景对话 Situational Dialogue

A: The party was **awesome**⑥ yesterday. I haven't had so much fun for quite a while now.

B: Sure thing. But yesterday you guys ordered too many meat dishes, especially crayfish. My friend is a vegetarian.

A: Oh no, why didn't she say so? Anyway, the Chinese always say, "There's no pleasure without meat." What meaning is that if there's no meat on the dining table?

B: Once, I joined her and her vegetarian friend for a meal. All the dishes were vegetables, tofu and so on. She also explained to me many benefits of going on a vegetarian diet.

A: I feel that vegetarians are **picky**[7] eaters, so what benefits can possibly be?

B: You can't say that. First of all, there're different categories of vegetarians. Some are vegetarians because of religious beliefs, for instance, it's only natural that **Buddhists**[8] cannot eat meat; some are vegetarians for the sake of protecting animals. What does that have to do with being picky eaters?

A: And your friend?

B: She did it for the sake of health. She said that she would be healthier and more relaxed after eating vegetarian food. Moreover, vegetarianism is also beneficial to the environment.

A: And what has this got to do with the environment?

B: According to her account, livestock such as cows and sheep give off gas which pollutes the environment. Excessive **consumption**[9] of beef or mutton will increase emission of such gas, and this adds to environmental pollution. So, eating less meat is good for the environment.

A: Really? No wonder more and more people are becoming vegetarians.

A: 昨天的聚会太棒了,我很久没有这么开心了。

B：是啊。不过，昨天你们点的肉菜太多了，尤其是虾。我的朋友吃素。

A：噢，她怎么不早说呢？不过，中国人总是说"无肉不欢"，如果餐桌上没有肉又有什么意思呢？

B：有一次，我和她还有她的素食朋友一起吃饭。所有的菜都是蔬菜、豆腐等。她还向我解释了素食的诸多好处。

A：我觉得素食者都很挑食，吃素能有什么好处呢？

B：你不能这么说。首先，有不同种类的素食者。有些人因为宗教信仰而吃素，比如，佛教徒不能吃肉是很自然的；有些人为了保护动物而吃素。这跟挑食有什么关系呢？

A：那你的朋友呢？

B：她是为了健康才这么做的，她说吃素后身体更健康、更放松。而且吃素对环境也有好处。

A：这和环境有什么关系？

B：听她说牛、羊等牲畜排放的气体会污染环境。过度食用牛羊肉会增加这种气体的排放，增加环境污染。所以，少吃肉对环境有好处。

A：真的吗？难怪越来越多的人正成为素食者。

 生词注解 Notes

① immerse /ɪˈmɜːs/　*vt.* 浸；沉湎于……

② crayfish /ˈkreɪfɪʃ/　*n.* 小龙虾；淡水龙虾

③ discrimination /dɪˌskrɪmɪˈneɪʃn/　*n.* 歧视；区别

④ barrier /ˈbærɪə/   n. 障碍

⑤ dominate /ˈdɒmɪneɪt/   vt. 支配；占优势

⑥ awesome /ˈɔːsəm/   adj. 令人敬畏的；使人畏惧的

⑦ picky /ˈpɪkɪ/   adj. 挑剔的；吹毛求疵的

⑧ Buddhist /ˈbʊdɪst/   n. 佛教徒

⑨ consumption /kənˈsʌmpʃən/   n. 消费；消耗

糖油粑粑

Sugar Oil Baba

 导入语 Lead-in

糖油粑粑是湖南长沙的地方传统名小吃。糖油粑粑造价低,主要原料是水磨糯米粉和糖汁,但其制造工艺特殊、精细讲究,圆溜油亮,黄而不焦,软而不粘,香甜爽口,虽然难登大雅之堂,更无法与山珍海味相媲美,但正是因为价格低廉,所以才能出入平常百姓家,受到大家喜爱,成为民间久吃不厌的小吃。长沙人都有吃糖油粑粑的美妙记忆和特殊情结。

饮食文化

文化剪影　**Cultural Outline**

"Baba" is a local **dialect**① of Changsha, which means cake. Sugar Oil Baba is made of glutinous rice flour and sugar juice, which is a noted traditional local snack in Changsha, Hunan Province. Its cost is cheap, but the **manufacturing**② process is exquisite.

"粑粑"是长沙方言，意思是饼。糖油粑粑由糯米粉和糖汁制成，是湖南长沙的地方传统小吃，虽然造价低廉，但制作工艺精细讲究。

Cuisine represents an attitude towards life for us now. Changsha people have a special liking for "Baba". Don't **underestimate**③ Sugar Oil Baba, for it plays an important role in Changsha people's life.

当下，美食之于我们代表了一种生活态度。长沙人对"粑粑"情有独钟。别小看糖油粑粑，它在长沙人的美食生活中占有重要地位。

Enjoying delicacies is an indispensable part for people who love life. The sweetness of Sugar Oil Baba make people feel happy in their heart and life.

对于热爱生活的人来说，享受美食是生活中不可缺少的部分。糖油粑粑不仅甜到人们的心里，也甜到幸福的生活里。

佳句点睛 Punchlines

1. China's public health **attainment**④ and **nutrition**⑤ knowledge has gradually improved.

中国公众健康素养和营养知识水平已经逐渐提高。

2. Healthy eating habits can promote the normal operation of human body and improve **immunity**⑥.

健康饮食习惯可以促进人体机能的正常运转,提高免疫力。

3. Sugar Oil Baba is golden, crisp and tender, and has an attractive color, making your mouth water.

糖油粑粑金黄脆嫩,色香诱人,让人垂涎欲滴。

情景对话 Situational Dialogue

A: What would you like for dessert?

B: What do you have?

A: I have Sugar Oil Baba, ice cream, chocolate cake and fruit **cocktail**⑦.

B: Can I have Sugar Oil Baba with ice cream?

A: Of course. I made Sugar Oil Baba this morning, so it's lovely and fresh.

饮食文化

B: I love your home-made Sugar Oil Baba. It's yummy. Can I have another glass of **lemonade**®?

A: Surely. You can get it yourself, it's in the refrigerator.

B: OK, would you like a drink too?

A: Yes, I'll have an cup of iced tea, just next to the lemonade.

B: Are you having any dessert?

A: I'll have Sugar Oil Baba too, but without ice cream.

B: Here you are.

A: Wow, how fantastic!

B: It's made of glutinous rice flour and sugar juice.

A: Well, no wonder it's so savory.

A: 你想要什么甜点?

B: 你这里有什么?

A: 我这里有糖油粑粑、冰淇淋、巧克力蛋糕和水果鸡尾酒。

B: 我能要糖油粑粑和冰淇淋吗?

A: 当然能。我今早刚做的,很新鲜、很好吃。

B: 我喜欢你自制的糖油粑粑,很好吃。能再给我来杯柠檬水吗?

A: 当然能。你自己拿吧,在冰箱里。

B: 好的。你想要喝点儿什么?

A: 我想要杯冰茶,就在柠檬水旁边。

B: 你要甜点吗?

A: 我也要糖油粑粑,别加冰淇淋。

B: 给你。

A: 哇,味道好极了。

B: 这是由糯米粉和糖汁做成的。

A: 嗯,难怪这么好吃。

生词注解　Notes

① dialect /ˈdaɪəlekt/　n. 方言;土话

② manufacturing /ˌmænjuˈfæktʃərɪŋ/　n. 生产;制造

③ underestimate /ˌʌndərˈestɪmeɪt/　vt. 低估;看轻

④ attainment /əˈtenmənt/　n. (常用复数)造诣;学识

⑤ nutrition /njuˈtrɪʃn/　n. 营养

⑥ immunity /ɪˈmjuːnəti/　n. 免疫力;豁免权

⑦ cocktail /ˈkɒkteɪl/　n. 鸡尾酒;开胃食品

⑧ lemonade /ˌleməˈneɪd/　n. 柠檬水

桂林米粉

Guilin Rice Noodles

 导入语 Lead-in

广西桂林米粉,也称为"卤肉粉",首次出现于唐朝,改良于宋朝,经过长期的地域演变,在云南演变为过桥米线,在贵州变成了肠旺粉,在广东变成了炒河粉,在广西柳州成了螺蛳粉和卷粉。即使在桂林,米粉也有汤粉和炒粉之分,但最正宗的桂林米粉还是卤粉。主料有干米粉、五香牛肉、骨头汤、酸豆角、笋尖、花生米,配料有小青菜、盐、辣椒油。在桂北地区,米粉也代替了绣球充当传情达意的功能,如果男方看上了某个姑娘,他就会当着媒人的面请姑娘去吃米粉,如果姑娘欣然接受,那就表明姑娘对男方有意,这门亲事就算定下来了,这就是祖辈们常说的"一碗米粉定终身"。

文化剪影 Cultural Outline

Guilin Rice Noodles is mainly made up of rice noodles, **brine**[①] and pickled vegetables. The most authentic Guilin Rice Noodles attracts diners from all over the world with their characteristics of tenderness, freshness and tastiness.

桂林米粉主要由米粉、卤水和卤菜组成。最正宗的桂林米粉以其嫩、鲜、香的特色吸引了八方食客。

Guilin Rice Noodles is the **originator**[②] of fast food in the world and the representative of Chinese cuisine culture. The rice noodles suitable for all seasons has become the main food for Guilin people and the must-order table for tourists from all over the world.

桂林米粉是世界快餐业的鼻祖和中国饮食文化的代表。四季皆宜的米粉成为桂林人的主食和世界各地游客光临桂林必点的佳肴。

The key to Guilin Rice Noodles is to mix them with a strong and tasty brine, which is boiled with spices and traditional Chinese medicine according to a certain **formula**[③], which is also **favored**[④] by most people for its a strong **efficacy**[⑤] of traditional Chinese medicine.

桂林米粉美味的关键是要与浓而又香的卤水混合。卤水是将香料和中药按照一定的配方熬制而成,具有较强的中药功效,因此也受到人们的青睐。

佳句点睛 Punchlines

1. Guilin Rice Noodles is a famous snack in China, which can be **rated**⑥ as one of the best snacks in Guilin.

桂林米粉是中国著名小吃,堪称桂林小吃一绝。

2. Guilin Rice Noodles is a classic of cooking art because of its unique production method.

桂林米粉制作方法独特,是烹调艺术的经典。

3. Guilin Rice Noodles not only stands for a snack, but also witnesses of ethnic integration.

桂林米粉不仅是小吃,也是民族融合的见证。

 情景对话 Situational Dialogue

A: Are you hungry now?

B: No. Are you **kidding**⑦ me? We just ate some cookies an hour ago.

A: That is not my fault. All the food smells so savory today and makes my mouth water!

B: Are you serious?

A: You know what? I guess it is so cold right now that a **gust**⑧ of strong wind has taken all calories from me. I need to have something to eat.

B: You win!

A: I'd like to have rice noodles!

B: OK, there's a famous rice noodle restaurant. Let's go.

A: Here is the menu outside.

B: What's on the menu?

A: There're many different kinds of rice noodles.

B: I just order Guilin Rice Noodles.

A：你现在饿吗？

B：不饿。你是在逗我吗？我们一小时前刚吃了饼干。

A：那不是我的错。今天满街的食物都闻着好香，我都流口水了！

B：你当真？

A：你知道吗？我觉得是太冷了，强风把我的卡路里都带走了。我要吃些东西。

B：你赢了。

A：我想吃米粉。

B：行，那里有一家很有名的米粉馆。咱们走吧！

A：外面有菜单。

B：菜单上都有什么？

A：很多种不同米粉。

B：我就要桂林米粉。

生词注解 Notes

① brine /braɪn/ n. 卤水；盐水

② originator /əˈrɪdʒɪneɪtə(r)/ n. 创始人；发明者

③ formula /ˈfɔːmjələ/ n. 配方；处方

④ favor /ˈfeɪvə(r)/ vt. 偏爱；赞许

⑤ efficacy /ˈefɪkəsɪ/ n. 功效；效力

⑥ rate /reɪt/ vt. 评价；估价

⑦ kid /kɪd/ vt. 戏弄；取笑

⑧ gust /ɡʌst/ n. 突然一阵

螺蛳粉

Liuzhou River Snails Rice Noodles

导入语 Lead-in

螺蛳粉是广西柳州地区的一种传统名小吃,最早起源于20世纪70年代的柳州夜市,具有辣、爽、酸、烫的独特风味。螺蛳粉主料有螺蛳、米粉、酸笋、青菜、花生、木耳、腐竹、黄花菜、萝卜干、辣椒油,配料有山柰、八角、肉桂、丁香等。螺蛳粉之所以让人欲罢不能,与它的色、香、味密不可分。首先是独有的配料"腌制酸笋"。作为腌制发酵食品,发酵过程中自然会产生一些氨基酸物质,于是就会有酸味。经过腌制发酵,其特殊的"酸臭味"使螺蛳粉变得回味无穷。2020年,螺蛳粉被列入《第五批国家级非物质文化遗产名录》。

 文化剪影　Cultural Outline

Liuzhou Municipal Government began to promote the supply-side structural reform, **formulated**[①] the production standard of bagged Liuzhou **River Snails**[②] Rice Noodles in time, and introduced **relevant**[③] policies for the development of river snail noodle products to improve the supply **efficiency**[④].

柳州市政府开始推动供给侧结构性改革，及时制定袋装螺蛳粉生产标准，出台螺蛳粉产品发展相关政策，提高供给效率。

There's a pet phrase in Liuzhou that "people would rather eat without meat, not without River Snails Rice Noodles." It can be seen that River Snails Rice Noodles have become Liuzhou's most local characteristics of the famous snacks.

柳州人有句口头禅："宁可食无肉，不可无螺蛳粉"。由此可见，螺蛳粉已经成为柳州最具地方特色的著名小吃。

The reason why Liuzhou River Snails Rice Noodles is called because its soup is boiled with snails. The base of Liuzhou River Snail Rice Noodle Soup has a unique taste after being boiled with snail meat.

螺蛳粉之所以得名，是因为其汤是由螺蛳熬成的。螺蛳粉的汤底经过螺肉熬制，会散发出一股独特的味道。

佳句点睛 Punchlines

1. Liuzhou River Snails Rice Noodles has become a delish cyber **celebrity**⑤ food best sold around the world.

螺蛳粉已经成为畅销全球的网红美食。

2. Liuzhou River Snails Rice Noodles attracts many diners as the taste and color are good.

螺蛳粉因色味俱佳而吸引了众多食客。

3. Liuzhou River Snails Rice Noodles has a unique taste and is rich **ingredients**⑥.

螺蛳粉味道独特,配料丰富。

情景对话 Situational Dialogue

A: Recently, I heard a piece of news that South Korean people like Liuzhou River Snails Rice Noodles very much and wanna apply for the world **intangible**⑦ cultural **heritage**⑧ as South Korean food.

B: Yes, it was a hot topic some time ago.

A: Liuzhou River Snails Rice Noodles is a famous traditional snack with a history of half a century, and we certainly do not want to let South Korea apply first.

B: Yes, it will always belong to China and it is also gonna apply for the world intangible cultural heritage.

A: The South Korean people are so fond of it, which serves to show how delicious it is.

B: It has been developing rapidly in recent years and has become the **cyber**⑨ celebrity snack with the highest sales volume of Alibaba rice noodles.

A: The sour bamboo **shoots**⑩ in it, even though the smell is not good, but the more you have, the better taste.

B: Yes, together with the special sour smell of sour bamboo shoots plus the stimulation of chilli, it's hard to resist the food temptation.

A: It is really unique and suitable for the whole world!

B: Now the enterprises also wanna expand the overseas market, and more and more overseas friends will like it.

A: Let's go and have a bowl of Liuzhou River Snails Rice Noodles.

B: OK, let's go.

A: 最近,我听到一则消息说韩国人非常喜欢螺蛳粉,想为螺蛳粉申请世界非物质文化遗产。

B: 是的,这是前段时间的热门话题。

A: 螺蛳粉作为我国拥有半个世纪历史的著名传统特色小吃,我们当然不愿让韩国先申请。

B: 螺蛳粉是中国的,目前中国也正准备申请世界非物质文化遗产。

A：韩国人民都这么喜欢螺蛳粉，足见螺蛳粉到底有多好吃了。

B：螺蛳粉近几年发展迅速，成为阿里巴巴米粉特产类销售量排名第一的"网红"小吃。

A：螺蛳粉中的酸笋，闻起来臭，却越吃越香。

B：是的，酸笋的特殊酸臭加上辣椒的刺激，令人难以抗拒。

A：螺蛳粉真是很有特色，适合推向全世界！

B：现在螺蛳粉企业也要拓展海外市场，会有越来越多海外友人喜欢的。

A：走，我们也去吃碗螺蛳粉吧。

B：行，咱们走吧。

 生词注解 Notes

① formulate /ˈfɔːmjuleɪt/ vt. 规划；明确表达

② river snail /ˈrɪvə(r)sneɪl/ n. 田螺

③ relevant /ˈreləvənt/ adj. 相关的；切题的

④ efficiency /ɪˈfɪʃnsɪ/ n. 效率；效能

⑤ celebrity /səˈlebrətɪ/ n. 名人；名声

⑥ ingredient /ɪnˈɡriːdɪənt/ n. 材料；佐料

⑦ intangible /ɪnˈtændʒəbl/ adj. 无形的；触摸不到的

⑧ heritage /ˈherɪtɪdʒ/ n. 遗产；传统

⑨ cyber /saɪbə(r)/ n. 网络

⑩ shoot /ʃuːt/ n. 新芽；嫩枝

过桥米线

Crossing Bridge Rice Noodles

 导入语 Lead-in

云南过桥米线源自滇南蒙自县，始于明末清初。南方米食的制作和吃法多种多样，称谓也花样百出。就云南而言，除了最普通的将米煮成饭的吃法外，米线最为流行。米线的制作方法种类繁多，有过桥米线、炒米线、凉米线、小锅米线等，其中又以过桥米线最为有名。过桥米线流传至今，其制作方法丰富精细，主料有猪里脊肉片、鸡脯肉片、乌鱼片、猪腰片、肚头片、水发鱿鱼片，配料有豌豆尖、韭菜、香菜、葱丝、姜丝、草芽丝、玉兰片、豆腐皮。过桥米线含有丰富的维生素、酵素和矿物质等。

文化剪影 Cultural Outline

Rice noodles can be cooked or fried in a **variety**① of ways, but Crossing Bridge Rice Noodles is rich in ingredients, fine production and unique eating.

米线的烹制方法多样,可煮也可炒,而过桥米线的过人之处在于配料丰富、制作精细、吃法独特。

Rice noodles is an indispensable food in the diet of Yunnan people. Because of its unique cultural **connotation**②, Crossing Bridge Rice Noodles has risen from a simple food to a popular diet with local cultural characteristics.

米线是云南人饮食中不可或缺的一种食物。因其融入了特有的文化内涵,过桥米线已经从一道简单的美食上升为一种极具地方文化特色的大众食物。

Crossing Bridge Rice Noodles is made up of rice noodles, soup and ingredients. The **soup-stock**③ is the specialty of Crossing Bridge Rice Noodles, and also one of the exquisite techniques of chefs.

过桥米线由米线、汤和配料组成。高汤是过桥米线的特色,也是厨师讲究的工艺之一。

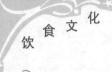

 佳句点睛　Punchlines

1. Crossing Bridge Rice Noodles has long been integrated into the food culture system of Yunnan **nationalities**④.

过桥米线早已融入云南各民族的饮食文化体系之中。

2. Crossing Bridge Rice Noodles tastes yummy, which is a kind of nutrition food.

过桥米线口味鲜美,是一种营养食品。

3. Crossing Bridge Rice Noodles also has a social function of **enlightening**⑤ the public.

过桥米线也具有教育启发大众的社会功能。

 情景对话　Situational Dialogue

A: Hey, what are you doing now?

B: I'm watching TV. Come on, it's really wonderful.

A: It's your **favorite**⑥ show?

B: Yes. I would like to watch food programs.

A: Why?

B: Because. it looks so tempting and tastes so delish.

A: That sounds good. What are they introducing now?

B: They are introducing Cross Bridge Rice Noodles for us.

A: Ahh, I think it's a good way to practice the chopsticks. You know foreigners are not good at using chopsticks.

B: That's right, we can practice it after making Cross Bridge Rice Noodles.

A: If you do it well, you'll make your Chinese friends quite an **impression**[⑦].

B: Yes, we need more practice.

A: 嘿,你在干什么呢?

B: 我在看电视。一起看吧,真的很棒。

A: 这是你最喜欢的节目?

B: 是的,我喜欢食品类节目。

A: 为什么呢?

B: 因为食物太诱人了,吃起来也很美味。

A: 听起来不错。他们在介绍什么?

B: 过桥米线。

A: 啊,我认为这是练习筷子的好方法。你知道,外国人不擅长使用筷子。

B: 是的,我们可以在做完过桥米线后练习。

A: 如果你做得好,会给你的中国朋友留下深刻印象。

B: 对,我们需要更多的练习。

 生词注解 Notes

① variety /vəˈraɪətɪ/ n. 多样；种类

② connotation /ˌkɒnəˈteɪʃn/ n. 内涵

③ soup-stock /suːpˈstɒk/ n. 高汤；老汤

④ nationality /ˌnæʃəˈnælətɪ/ n. 国籍；民族

⑤ enlighten /ɪnˈlaɪtn/ vt. 启发；启蒙

⑥ favorite /ˈfeɪvərɪt/ adj. 最喜欢的

⑦ impression /ɪmˈpreʃn/ n. 印象；效果

鲜花饼

Flower Cake

导入语 Lead-in

鲜花饼是以云南特有的食用玫瑰花入料的酥饼,是具有云南特色的招牌点心。鲜花饼口感酥软,香甜不腻,带有淡淡的花香。云南食用鲜花饼始于清代,工艺考究,采摘含苞欲放或刚刚开放的食用鲜花,经筛选后去掉花托、分开花瓣,加工处理为馅心,然后包上酥皮、烘烤冷却后成型。鲜花饼的制作过程要求及时迅速,以充分保留鲜花饼的鲜香。鲜花饼的主料有面粉、玫瑰、玉兰、菊花、冰糖、白糖、芝麻、花生、核桃仁、猪油。鲜花饼种类丰富,尤以玫瑰花饼见长,玫瑰花陷包括新鲜玫瑰花、蜂蜜、熟粉。可以说,鲜花饼既是美容保健的佳品,也是天然健康的补品。

饮食文化

 文化剪影　Cultural Outline

Flower Cake is made of **edible**① roses in **bud**② and prepared by **baking**③ with **refined**④ flour, sugar, honey and other ingredients.

鲜花饼采用含苞欲放的食用玫瑰制成玫瑰花馅,加以精制面粉、白糖、蜂蜜等配料烘烤而成。

Flower Cake is the classic representative of Yunnan mooncakes in China. The raw materials and techniques of flower cakes are technologically **innovative**⑤ on the basis of maintaining a good tradition.

鲜花饼是中国滇式月饼的经典代表。鲜花饼的原料、工艺在保持优良传统的基础上进行了技术创新。

The flower foods are beautiful in color, rich in **protein**⑥, various **amino acids**⑦ and **vitamins**⑧, which meets the needs of people's diets to return to nature and health. Therefore, the market scale is developing rapidly.

鲜花食品色彩美丽,富含蛋白质、多种氨基酸和维生素,符合人们饮食回归自然和健康的需求。因此,市场规模发展迅速。

 佳句点睛　Punchlines

1. With the improvement of people's living standards and the cog-

nition of fashion diet, flower foods are increasingly **favored**① by people.

随着人们生活水平的提高,以及对时尚饮食的认知,鲜花食品也越来越受到人们的青睐。

2. Roses are the most important factor in the preparation of fillings, and the roses themselves have a bitter taste due to the **tannins**⑪ contained in the roses.

玫瑰花是馅料制备中最重要的因素,其中含有的单宁酸使玫瑰花本身具有苦涩味。

3. The amount of roses added will have a great influence on the taste and flavor of the flower cakes.

玫瑰花的添加量会对鲜花饼的口感、风味产生极大影响。

情景对话　Situational Dialogue

A: I want something sweet after dinner.

B: What do you have in mind?

A: A dessert sounds nice.

B: What kind do you wanna get?

A: I wanna get some cake.

B: What kind of cake do you want?

A: I have no idea.

B: Do you want to know what kind of cake I like?

A: Surely. What kind do you like?

B: I love Flower Cake.

A: Flower Cake? Can flowers be eaten?

B: Of course, it's very delicious.

A: Wow, I'll have a good taste.

A: 晚饭后我想吃点儿甜的。

B: 你有什么想法?

A: 甜点听起来不错。

B: 你想买什么样的?

A: 我想买些甜饼。

B: 你想要什么样的甜饼?

A: 我不知道。

B: 你想知道我喜欢什么样的甜饼吗?

A: 当然想,你想要哪一种?

B: 我喜欢鲜花饼。

A: 鲜花饼? 鲜花可以吃吗?

B: 当然可以,非常好吃。

A: 哇,我得好好尝尝。

 生词注解 Notes

① edible /ˈedəbl/ *adj.* 可食用的

② bud /bʌd/ n. 蓓蕾

③ baking /ˈbeɪkɪŋ/ n. (在烤炉里)烘烤;烘焙

④ refined /rɪˈfaɪnd/ adj. 去掉杂质的;精炼的

⑤ innovative /ˈɪnəveɪtɪv/ n. 创新的;新颖的

⑥ protein /ˈprəʊtiːn/ n. 蛋白质

⑦ amino acid /əˈmiːnəʊ ˈæsɪd/ n. 氨基酸

⑧ vitamin /ˈvɪtəmɪn/ n. 维生素

⑨ favor /ˈfeɪvə(r)/ n. 支持;优惠

⑩ tannin /ˈtænɪnz/ n. 单宁酸;鞣酸

粤式早茶

Guangdong Morning Tea

导入语 Lead-in

粤式早茶的历史可以追溯到清朝咸丰、同治年间，后来出现了茶居，逐渐演变成茶楼，此后广东人去茶楼喝茶蔚然成风。早茶习俗多见于中国南方地区，尤其是广东和江苏地区。"早茶"并不等于饮茶，实质上是到酒楼"吃早餐"。广东人喜欢称之为"叹早茶"，"叹"在广东话中有"享受"的意思，由此可见喝早茶在广东人心中是一种愉快的消遣。茶点分为干、湿两种：干点有饺子、粉果、包子、酥点等，湿点有粥类、肉类、龟苓膏、豆腐花等，其中又以干点最为精致。粤式早茶以肠粉、叉烧包、酥皮菠萝包、虾饺、烧麦、西关名点等为特色，常见的茶品有铁观音、普洱和菊花。

文化剪影 Cultural Outline

The Chinese people are one of the first nations to **cultivate**① and drink tea, which is enjoyed by people from all social classes. There're different varieties of tea **available**②, depending on the different producing areas.

中国人是最早种茶、饮茶的民族之一,茶深受社会各阶层人民的喜爱。根据产地的不同,茶叶的品种也有所不同。

Drinking tea is the most basic and proudest cuisine culture of Guangdong. Guangdong people drink tea three times a day in the morning, at noon and in the evening, while the morning tea is the most exquisite.

饮茶是广东人最基础的也是最引以为豪的饮食文化。广东人每日早、中、晚三次品茶,其中早茶最为讲究。

It is a tradition for Guangdong people to go to the teahouse to drink tea in the morning. No matter it is a gathering of family or friends, they always like to go to tea house to **brew**③ a pot of tea and order some snacks.

广东人早晨去茶楼喝茶是一种传统,无论是家人聚会还是朋友见面,都爱去茶楼,泡上一壶茶,点上几份点心。

饮食文化

 ## 佳句点睛 Punchlines

1. China is the birthplace of tea, with a long history of drinking tea and **profound**④ tea culture.

中国是茶的发源地,具有悠久的饮茶历史和深厚的茶文化底蕴。

2. Tea is commonly enjoyed with dim sum or **desserts**⑤, but can also be taken after a meal.

茶通常与点心或甜品一起享用,但也可以在餐后享用。

3. The everlasting **prosperity**⑥ of Guangdong Morning Tea is inseparable from the prosperity of trade and the rapid economic development in Guangdong since ancient times.

广东早茶的长盛不衰,与广东自古以来贸易兴盛和经济快速发展密不可分。

 ## 情景对话 Situational Dialogue

A: Good morning. What can I help you?

B: I would like something to drink.

A: What kind of drinks would you like?

B: Do you have any teas?

A: Of course, we have lots of teas.

B: What do you recommend?

A: What about green tea or perhaps **jasmine**① tea?

B: What's this one?

A: That's Oolong Tea.

B: OK, I'll try this.

A: Our cakes taste very well. You could have a try. There're a lot of choices, such as honey raisin **scone**②, marble cheesecake, green tea **tiramisu**③ and so on. Besides, we have promoted sweet potato cupcake recently.

B: Well, give me a sweet potato cupcake, please.

A: 你好,请问需要什么吗?

B: 我想喝点饮品。

A: 你想喝点什么?

B: 你们有茶吗?

A: 当然,我们有很多茶。

B: 你有什么好的推荐吗?

A: 绿茶或茉莉花茶怎么样?

B: 这是什么?

A: 这是乌龙茶。

B: 好的,我试试这个。

A: 我们的蛋糕很好吃。你可以尝尝。有很多选择,如蜂蜜提子司康、大理石芝士蛋糕和抹茶提拉米苏等。我们最近还新推出了甘薯杯子蛋糕。

B: 好, 请给我来一块甘薯杯子蛋糕吧。

生词注解 Notes

① cultivate /ˈkʌltɪveɪt/ vt. 培养;陶冶

② available /əˈveɪləbl/ adj. 可用的;可获得的

③ brew /bruː/ vt. 酿造;泡茶

④ profound /prəˈfaʊnd/ adj. 深厚的;意义深远的

⑤ dessert /dɪˈzɜːt/ n. (餐后)甜食;甜点

⑥ prosperity /prɒˈsperətɪ/ n. 繁荣;成功

⑦ jasmine /ˈdʒæzmɪn/ n. 茉莉;淡黄色

⑧ scone /skɒn/ n. 烤饼;司康饼

⑨ Tiramisu /ˌtɪrəməˈsuː/ n. 提拉米苏(一种意大利式甜点)

煲仔饭

Steamed Rice in Clay Pot

导入语 Lead-in

煲仔饭，也称"瓦煲饭"，是一种广东传统美食。将淘好的米放入煲中，把米饭煲至七成熟，加入配料，再转用文火煲熟即可。煲仔饭的传统种类主要有豆豉排骨饭、腊味饭、香菇滑鸡饭、猪肝饭、烧鸭饭、白切鸡、黄鳝饭等。制作煲仔饭须用砂锅，砂锅具有受热均匀与耐高温的特性，这样可以更好地保存食物养分，煲出来的饭也特别香。煲仔饭主料是大米，最适合煲仔饭的是形状细长的丝苗米，煲出的饭颗粒分明、油润晶莹，满口余香，回味悠长。

 文化剪影 Cultural Outline

Steamed Rice in Clay Pot looks simple, but if one wants to make a bowl of **superb**① steamed rice in clay pot, he should identify the quality of materials, hold the fire strictly and cook the food properly.

煲仔饭看来做法简单,但要做得好,就要甄别各种用料的品质,严格把握火候,做到恰到好处。

Rice, is the **staple**② food in southern China. It can be mixed with five flavors and almost supply all the **nutrients**③ the whole body needs.

米饭,是中国南方的主食,可以与五味调配,几乎可以供给身体所需的全部营养。

The **sauce**④ is the essence of Steamed Rice in Clay Pot. The seasoning is very simple while in making process it is cooked without any seasoning, so the flavor of sauce is very important.

煲仔饭的精髓是料汁,煲仔饭的调味很简单,在煲制时不加任何调料,因此料汁的味道非常重要。

 佳句点睛 Punchlines

1. Steamed Rice in Clay Pot is one of the representative special

dishes of Guangdong Province.

煲仔饭是广东省代表性的特色菜肴之一。

2. The materials of Steamed Rice in Clay Pot are **flexible**⑤ and **diverse**⑥, and can be different according to time and place.

煲仔饭的用料灵活多样，可以因时、因地而异。

3. Rice food not only represents the meaning of cuisine, but also symbolizes the **inheritance**⑦ of Chinese culture.

米食不仅代表饮食的含义，而且象征中华文化的传承。

 情景对话 **Situational Dialogue**

A: Can you tell me where I can find a restaurant not too expensive?

B: As far as I can see, Guangzhou Restaurant is a good choice.

A: What is its specialty?

B: Steamed Rice in Clay Pot there is very nice.

A: Steamed Rice in Clay Pot?

B: Steamed Rice in Clay Pot belongs to Guangdong Cuisine. The main types of clay pots are sausage clay pot, mushroom clay pot, chicken clay pot, roast duck clay pot, and so on, among which the major ingredient is rice.

A: Rice is really important in the Chinese diet.

B: Yeah! The main component of rice is **carbohydrate**⑧, its nutritional value complete and balanced.

A: How can I get there?

B: It's right across the street.

A: Thank you.

B: My pleasure.

A: 你能告诉我在哪里可以找到不太贵的饭店吗？

B: 据我所知，广州饭店不错。

A: 都有什么特色菜？

B: 那里的煲仔饭很不错。

A: 煲仔饭？

B: 煲仔饭，属于粤菜。主要有香肠煲、香菇煲、鸡煲和烤鸭煲等，主料是米。

A: 米在中国饮食中真重要。

B: 是啊！米的主要成分是碳水化合物，它的营养价值完整均衡。

A: 我怎么去那里呢？

B: 就在街对面。

A: 谢谢你。

B: 不客气。

生词注解 Notes

① superb /suːˈpɜːb/ *adj.* 极好的；华丽的

② staple /ˈsteɪpl/ *adj.* 主要的；大宗生产的

③ nutrient /ˈnjuːtrɪənt/ *n.* 营养素

④ sauce /sɔːs/ *n.* 酱油；调味汁

⑤ flexible /ˈfleksəbl/ *adj.* 灵活的；柔韧的

⑥ diverse /daɪˈvɜːs/ *adj.* 多种多样的

⑦ inheritance /ɪnˈherɪtəns/ *n.* 继承；遗传

⑧ carbohydrate /ˌkɑːbəʊˈhaɪdreɪt/ *n.* 碳水化合物

云吞面

Wonton Noodles

导入语 Lead-in

云吞面,又称为"馄饨面""细蓉""大蓉",是广东省的特色名小吃。广东人爱吃云吞面,一碗上乘的云吞面有三讲:一讲面,必须是竹升打的银丝面;二讲云吞,要三七开肥瘦的猪肉,是云吞馅的最佳比例,还要用鸡蛋黄浆住肉味;三讲汤,要大地鱼、虾籽、猪骨、火腿熬成的浓汤。面条劲道爽口、用料扎实、精工细作。不放一滴水,只加入鸡蛋,手工和面,再依靠人力骑在竹竿上弹跳,一边压打,一边滚动,如此反复一两个小时后,面团就能揉拉成一根根细如银丝的面条。

文化剪影 Cultural Outline

Wonton Noodles is a Guangdong noodle dish which is popular in Guangzhou and Hong Kong of China, as well as Malaysia, Singapore and Thailand. The dish is usually served in a hot **broth**①, **garnished**② with leafy vegetables and Wonton Noodles.

云吞面是一种广式面食，在中国广州、香港，以及马来西亚、新加坡和泰国都很受欢迎。这道美食通常是在热汤中点缀着蔬菜和馄饨。

Noodles are not a staple food in Guangdong, but an authentic bowl of Wonton Noodles carries not only delicacies but also contains the memory of hometoun.

面食在广东并不是主食，但一碗地道的云吞面承载着的不仅仅是美味佳肴，而且包含着乡土的记忆。

Wonton Noodles is one of the authentic Guangdong snacks, among which there're plenty of **variations**③ of this popular Guangdong dish, including different **toppings**④ and garnishes. For example, the noodles are served relatively dry or the noodles dipped in the soup for eating.

云吞面是一种地道的广东小吃，这道广受欢迎的粤菜有很多不同的做法，包括不同的配料和配菜。例如，汤和面分开吃或把面浸在汤里吃。

235

饮食文化

 佳句点睛 Punchlines

1. Wonton Noodles is one of the special snacks in Guangdong Province.

云吞面是广东的特色小吃之一。

2. Wonton Noodles is easy to digest and absorb and have the effect of improving **anaemia**⑤ and enhancing immunity.

云吞面易于消化吸收,有改善贫血、增强免疫力等功效。

3. Most of Guangdong people have an inseparable passion for Wonton Noodles.

大部分广东人对云吞面有一种难以割舍的情怀。

 情景对话 Situational Dialogue

A: If you go to Guangzhou, there's a snack deserving a try — that's Wonton Noodles.

B: I know it, but in a city that's famous for delicacies, how do you make your Wonton Noodles stand out?

A: There're two secrets: First, the broth must be out of the ordinary.

B: What does that mean?

A: The broth should be rich, yet light and **refreshing**⑥.

B: What about the second?

A: And then the noodles. The best chefs mix fresh eggs into their dough, which makes the noodles so **al dente**⑦.

B: Sounds great!

A: There're many features for Wonton Noodles, some of which are mainly **prawns**⑧, with small amounts of minced pork, or no pork at all.

B: How about the noodles?

A: The noodles are al dente, free from the taste and odor which is characteristic in many egg noodles when cooked.

B: I'm sure I've found the perfect snack.

A: 如果你去广州,有一种小吃值得一试——那就是云吞面。

B: 我知道,但在一座以美食闻名的城市,怎么才能让一碗云吞面脱颖而出呢?

A: 有两个秘诀:首先,肉汤必须与众不同。

B: 什么意思?

A: 肉汤应该味道浓郁,但又清淡爽口。

B: 第二个呢?

A: 然后是面条。最好的厨师会把新鲜鸡蛋拌到面团里,这样面条才会有嚼劲。

B: 听起来太棒了!

A: 云吞面特色各异,有的以大虾为主,有的放少量肉末,或者根本不放猪肉。

　　B: 面条怎么样？

　　A: 面条很有嚼头，没有许多鸡蛋面煮熟后特有的口感或气味。

　　B: 我敢说我找到了最完美的小吃。

 生词注解　Notes

① broth /brɒθ/　*n.* 肉汤；蔬菜清汤

② garnish /ˈgɑːnɪʃ/　*vt.* 装饰（尤指食物）；加饰菜于……

③ variation /ˌveərɪˈeɪʃən/　*n.* 变异；变化

④ toppings /ˈtɒpɪŋz/　*n.* 浇头；食品上的装饰配料

⑤ anaemia /əˈniːmɪə/　*n.* 贫血

⑥ refreshing /rɪˈfreʃɪŋ/　*adj.* 提神的；使……清爽的

⑦ al dente /ˌæl ˈdenteɪ/　*adj.*（尤指面食）筋道的；有韧性耐咀嚼的

⑧ prawn /prɔːn/　*n.* 对虾；明虾